Whitestein Series in Software Agent Technologies

Series Editors:
Marius Walliser
Stefan Brantschen
Monique Calisti
Thomas Hempfling

This series reports new developments in agent-based software technologies and agent-oriented software engineering methodologies, with particular emphasis on applications in various scientific and industrial areas. It includes research level monographs, polished notes arising from research and industrial projects, outstanding PhD theses, and proceedings of focused meetings and conferences. The series aims at promoting advanced research as well as at facilitating know-how transfer to industrial use.

About Whitestein Technologies

Whitestein Technologies AG was founded in 1999 with the mission to become a leading provider of advanced software agent technologies, products, solutions, and services for various applications and industries. Whitestein Technologies strongly believes that software agent technologies, in combination with other leading-edge technologies like web services and mobile wireless computing, will enable attractive opportunities for the design and the implementation of a new generation of distributed information systems and network infrastructures.

www.whitestein.com

Monique Calisti

An Agent-Based Approach for Coordinated Multi-Provider Service Provisioning

Springer Basel AG

Author's address:
Monique Calisti
Whitestein Technologies
Gotthardstrasse 50
8002 Zürich
Switzerland
e-mail: mca@whitestein.com

2000 Mathematical Subject Classification 68T35, 68U35, 94A99, 94C99

A CIP catalogue record for this book is available from the
Library of Congress, Washington D.C., USA

Bibliographic information published by Die Deutsche Bibliothek
Die Deutsche Bibliothek lists this publication in the Deutsche Nationalbibliografie;
detailed bibliographic data is available in the Internet at <http://dnb.ddb.de>.

ISBN 978-3-7643-6922-4   ISBN 978-3-0348-7972-9 (eBook)
DOI 10.1007/978-3-0348-7972-9

© 2003 Springer Basel AG
Originally published by Birkhäuser Verlag, Basel - Boston - Berlin in 2003

Cover design: Micha Lotrovsky, CH-4106 Therwil, Switzerland
Printed on acid-free paper produced from chlorine-free pulp. TCF ∞

ISBN 978-3-7643-6922-4

9 8 7 6 5 4 3 2 1   www.birkhauser-science. com

Contents

Abstract

Communication networks are very complex and interdependent systems requiring complicated management and control operations under strict resource and time constraints. A finite number of network components with limited capacities need to be shared for dynamically allocating a high number of traffic demands. Moreover, coordination of peer provider is required whenever these demands span domains controlled by distinct operators. This task is even more delicate for networks which aim to provide any kind of Quality of Service (QoS) guarantees. In this context, traditional human-driven management and control is becoming increasingly inadequate to cope with the growing heterogeneity of actors, services and technologies populating the current deregulated communication market.

This book proposes the Network Provider Interworking (NPI) paradigm as a novel approach to improve multi-provider interactions based on the coordination of autonomous and self-motivated software entities acting on behalf of distinct operators. Coordination is achieved by means of distributed constraint satisfaction techniques integrated within economic mechanisms, which enable automated negotiations to take place. This allows software agents to find efficient allocations of service demands spanning several networks without having to reveal strategic or confidential data. In addition, a novel way of addressing resource allocation and pricing in a compact framework is made possible by the novel use of powerful resource abstraction techniques.

The NPI approach is evaluated on a testbed simulating multi-provider network environments. Experimental results validate the new paradigm and demonstrate the feasibility of the developed techniques in realistic network scenarios. The comparison of the NPI-based framework to alternative techniques for automating the multi-provider interactions, such as providers coordination based either on static contracts (the FAS approach) or on coalitions centrally supervised by a meta-controller (the COB solution), confirms the benefits of a dynamic and distributed architecture. Finally, the analysis and the discussion of these results make it possible to identify the potential limits of this work.

Preface

The work presented in this volume has been developed during the years I spent at the Artificial Intelligence Laboratory of the Swiss Federal Institute of Technology of Lausanne (EPFL). This book represents a revised version of my PhD thesis.

I would like to thank all the people who have supported me throughout this work. Firstly, many thanks to Professor Boi Faltings, my academic supervisor at EPFL, for the opportunity he gave me to work on this topic and for many useful discussions that have been essential to overcome some of the most critical phases of this research activity. I am also particularly grateful to Marius Walliser and Stefan Brantschen not only because they gave me the opportunity to publish the results of five hard working years in this book, but most of all for the chance I got to join the Whitestein team.

I am very grateful to Professors Rachid Guerraoui, Jeff Bradshaw and Nick Jennings for their scientific and friendly support especially during these last months and, of course, for having accepted to read and judge this dissertation. Their feedback and suggestions have been precious for improving both the work.

A thesis is not only a scientific work and I would have not been able to get here without the great, unique and unconditional support of my family, Adrian and my great friends. Grazie di tutto!

Monique Calisti

Chapter 1

Introduction

The central topic of this volume is the coordination of network operators (or providers) for the allocation of service demands spanning distinct communication networks. Nowadays, the world-wide communication infrastructure is a network of domains controlled by autonomous, self-interested authorities and, therefore, carrying traffic across multiple domains usually requires cooperation - *inter-domain coordination* - of several operators. In particular, when traffic has to satisfy specific Quality of Service (QoS) requirements, this leads to even more complex negotiation problems since inter-domain coordination is complicated by the need to allocate and manage resources inside every network so that the required QoS can be guaranteed - *intra-domain* resource management and control.

This work addresses issues at both the inter-domain coordination and the intra-domain control levels by defining a set of techniques, which are the basis of an innovative paradigm for coordinating distinct network providers, while efficiently allocating resources inside every communication network. The term *Network Provider Interworking* (NPI) refers to the coordination framework proposed by this volume. Autonomous software entities, called *agents*, acting on behalf of different network operators coordinate their actions by means of distributed constraint satisfaction techniques integrated with economic mechanisms for automated negotiations.

The integration of distributed constraint satisfaction techniques allows agents to find efficient allocation of services spanning several networks without needing to reveal private constraints, i.e., strategic and confidential data. Automated negotiation methods and agent technology have been increasingly applied to the networks communication field. This trend is currently supported by important application pull and technology push factors. On one hand, there is a strong need for innovative solutions to cope with Telecom market liberalisation, rapidly changing networking technology [137], increasing flexibility and variety in usage requirements [71] and e-commerce needs and opportunities [299]. On the other hand, the increasingly dynamic network protocols, the availability of a large number of toolkits enabling developers to quickly build multi-agent systems, and the growth of standardisation efforts such as OMG [220] and FIPA [92], which provide standard interaction mechanisms for agent-based software, make agent application a real possibility.

This first chapter is set out as follows. We first motivate the work described in this volume by presenting the multi-provider coordination problem. Specific requirements at both inter- and intra-domain levels are considered and a simple example is discussed. Traditional approaches to multi-provider coordination are briefly revisited and the main principles upon which the innovative solution proposed by this work relies are introduced. The chapter goes on by introducing the main background techniques beyond the development of the NPI paradigm, namely multi-agent technology, automated negotiation, distributed constraint satisfaction methods and resource abstraction techniques. Finally, an outline of this volume concludes this chapter.

1.1   Research Motivations

The first aim of this work is to address the problem of maximising the distributed allocation of resources subject to incomplete information in highly dynamic environments. Within this general context, the specific solution proposed by our approach is a coordination paradigm to improve traditional multi-provider interactions for service set up. This work is motivated by both the pressing needs for more flexible and automated *interworking* solutions and the challenging problems for these solutions to be effective, feasible and usable in real environments. "Interworking" is used in this book to refer to the set of operations and processes involved in multi-provider interactions for the allocation of service demands spanning distinct networks.

In its most simple form, the multi-provider service setup problem is an optimisation (allocation) problem with assignment constraints. These constraints determine and influence the way a fixed amount of resources is allocated to a given number of activities needed to support the service in the most effective way (for good references about resource allocation problems and their variants see [136]). In the multi-provider context, both the increasing end user QoS requirements and the distribution of resources managed by distinct entities introduce challenging issues and additional constraints that make service allocation even more complex. Resource allocation problems have indeed been proven to be in general NP-complete [3].

In addition to the specific problems related to the multi-provider service set up process, the work described in this book is also motivated by the aim to develop efficient mechanisms that have the potential to be re-used in similar frameworks requiring coordination of software entities that, despite self-interests, need to exhibit cooperative behaviour to "improve their local performance" [179]. This can happen in several systems providing services such as e-banking, e-financing, air traffic control, production planning and control in the process industry, which directly build on top of distributed communication networks.

1.1.1 The Multi-Provider Coordination Problem

In the current deregulated communication market, the need to improve the way of allocating resources across distinct networks is becoming more and more urgent [312], [68]. On one hand, the increasing number of services demands requiring stringent QoS guarantees and spanning domains under the control of different operators (i.e., *multi-provider* service demands) pushes for better coordination. On the other hand, the heterogeneity of actors and technologies and the growing number of providers is augmenting the competition. Therefore, it becomes vital for every operator to define mechanisms that enable dynamic and efficient resource allocation according to growing interworking and coordination requirements [181], [336].

Currently, many aspects of the interactions between network providers are regulated by static long-term Service Level Agreements [331] (SLAs) and contracts that define the number and available capacity of links and network nodes (or access points) connecting one network domain to another, prices, etc., without taking into account the current network state. As a main consequence, intra-domain tasks and performance are strongly constrained, since the possibility to dynamically accommodate prices and balance the load on network resources is very limited or even not possible. This generates inefficient network resources utilisation with consequent profit losses [111].

In order to better illustrate the multi-provider coordination problem and the negative effects of uncoordinated and/or static policies that prevent taking into account the current state of the network, consider the following example. A typical service requiring the exchange of data between two remote points with specific QoS guarantees is video-conference. A video-conference is a live connection between people in separate locations for the purpose of communication, usually involving audio and often text as well as video. At its simplest, video-conferencing provides transmission of static images and text between two locations. At its most sophisticated stage, it provides transmission of full-motion video images and high-quality audio between multiple locations which requires minimal delay and cell loss at the network level. Consider the network scenario depicted in Figure 1.1 where the three communication networks (or network provider domains) X, Y, and Z

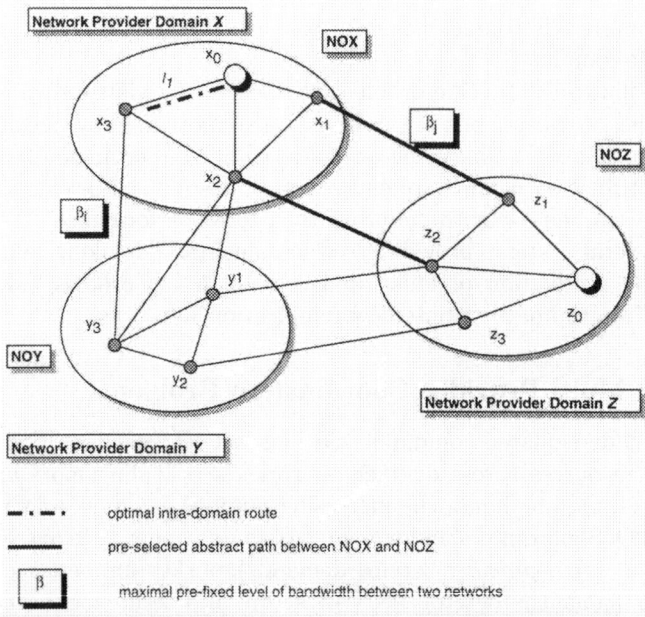

Figure 1.1: *A multi-provider network scenario consisting of three main communication networks controlled by distinct operators. Service demands spanning several networks require the coordination of different operators. When static agreements exist among them, the dynamic accommodation of resources is strongly limited or even impossible. This causes inefficiency and degradation in terms of network resource utilisation and profit.*

are controlled by three distinct network operators, namely NOX, NOY and NOZ. Assume that NOX receives a demand for establishing a video-conference between the two remote network nodes x_0 and z_0 requiring a certain amount of bandwidth β_k. At the inter-domain level, the provider NOX has to verify which peer operators it is possible to interact with in order to reach the final destination z_0, i.e., which possible abstract paths can be followed in order to interconnect the two remote nodes x_0 and z_0. An *abstract path* is an ordered list of distinct network provider domains between the source and the destination networks. In this example, there are in principle two possible alternative abstract paths: $X - Y - Z$ and $X - Z$. At the intra-domain level then, every operator involved in the video-conference setup has to select which specific network resources (i.e., nodes and links) the traffic will be routed through. This latter selection should be ideally based upon two main criteria [44]. First of all, only the internal routes that can support the required video-conference QoS requirements are considered, for instance, only the links with at least β_k available bandwidth are possible candidate routes. Second, among the set of possible internal routes, the final choice is based upon the optimisation of costs and network resources utilisation. Assume that, based on these criteria, the optimal choice for the operator NOX is represented by the local route x_0-l_1-x_3. If pre-fixed and static agreements establish that the traffic between the provider domains X and Z has to be routed along the abstract path $X - Z$ (with $\beta_j > \beta_k$) or that the inter-domain connections between NOX and NOY can only accommodate demands requiring at maximum β_i available bandwidth, with $\beta_i < \beta_k$, a different and sub-optimal route has to be searched instead of x_0-l_1-x_3. This generates sub-optimal resource allocation with consequent losses in terms of profit. In the worst case scenario, the incoming service demand might have to be rejected, even if the providers involved have enough resources to support it and a physical path between source and destination network nodes exists.

Abstracting from this concrete example to a more global and generic case, in this volume, the *Multi-provider Service Setup* (MuSS) problem is formulated as follows.

The Multi-provider Service Setup (MuSS) problem

Given a group of interconnected communication networks controlled by distinct authorities and consisting of nodes and bidirectional links, where each link has given resources capacities, and incoming multi-provider service demands allocations, each demand specified in terms of source and destination nodes, and a set of QoS requirements.

Find on-demand the end-to-end routes that satisfy connectivity constraints and QoS requirements over all distinct communication networks involved.

Select one end-to-end connection consisting of specific intra-domain routes and inter-domain links in coordination with all the other providers involved in the service demand allocation.

The expression 'on-demand' refers to the fact that the global end-to-end route is established when demand arises. In this case, the allocation process is state-based [44] since it takes into account the current network state information such as bandwidth availability, buffer occupancy, delays, etc. Therefore, suitable solutions to the MuSS problem seem to be *automated* frameworks supporting *dynamic* interactions between operators (see [18], [33], [351] and several works in [125]). This kind of approach has indeed the potential to provide a more flexible utilisation of network resources with a consequent reduction of costs and prices.

When defining a solution for the MuSS problem, the main challenges arise from the need to coordinate self-motivated entities that interact:

- Without having to reveal and centralise intra-domain topologies and internal strategic information.

- Having to make decisions based on structurally incomplete information.

- Dynamically taking into account the current network state (i.e., on-demand process).

- With fast and flexible mechanisms for peer-to-peer interactions.

- Having to integrate these software entities within existing networks.

This work is motivated by the aim of defining a solution to the MuSS problem that effectively meets these challenges. This requires action at two main levels.

First, at the *inter-domain* level, network operators need to dynamically coordinate themselves. This is achieved by exchanging the minimal amount of information necessary to define consistent end-to-end offers for final end customers. This information includes connectivity details about boundary network resources (e.g., which nodes and links will be selected for inter-domain routing) the offered QoS guarantees and prices. Concerning connectivity, for reasons of both scalability and confidentiality, providers do not reveal the details of their internal structure, but they disclose only a summary, or aggregated view, of the costs and availabilities of traversing their domains [91]. The need for aggregation is fundamental since computation and communication of routing protocols grows at least linearly the number of links in the network representation [11]. Concerning QoS guarantees and prices, current solutions widely rely upon static agreements that are pre-negotiated by human network administrators communicating with each other by telephone or facsimile. In the NPI framework, dynamic coordination, including information exchange, is achieved by combining distributed constraint satisfaction techniques with automated negotiations.

Second, at the *intra-domain* level, every provider has to efficiently allocate, manage, control and maintain its own network. These tasks are performed according to given end user requirements and profit maximisation objectives. More precisely, along the internal path selected for routing a specific incoming service demand, resources (i.e., bandwidth) should be optimally allocated to support the required QoS at a minimal cost. This generates the need for an efficient and flexible way of engineering the traffic [9] that is:

- Controlling and managing network elements, which also means an effective way of collecting network state information and keeping it up to date, i.e., network state control.

- Finding a feasible path based on the collected data, i.e., route computation.

- Pricing the set of resources, including operation and processes, required for supporting the service.

In this volume, a novel way of addressing network state control (i.e., bandwidth allocation), routing and pricing in a compact framework is enabled by the use of powerful clustering techniques. While traditional approaches to the multi-provider coordination problem are actually able to guarantee physical interconnectivity, service allocation and traffic routing in a quite static and human-dependent fashion (see Section 1.1.3), the solution proposed by this book can be considered as an evolutionary and innovative way of addressing both inter-domain and intra-domain issues listed above in a unified and automated framework (as anticipated in Section 1.1.4 and more exhaustively discussed in Chapters 4 and 5).

1.1.2   Why Bother about Bandwidth?

Whether mechanisms are needed to control and manage the allocation of bandwidth for guaranteeing QoS or not is a debated issue. There are two main opinions on that. One opinion is that fiber and dense wavelength division multiplexing (DWDM) technologies will make bandwidth so abundant and cheap that high quality of service will be automatically delivered without bothering too much about optimal allocation of existing resources. The other opinion is that no matter how much bandwidth and resources the Internet is able to provide, new applications will always be created to consume it. Therefore, ad hoc mechanisms to provide QoS will still be needed.

This latter argument represents an important motivation to this work. Even if bandwidth will eventually become abundant and cheap, this is not the case nowadays. Currently, communication networks are facing an increasing demand for a wider variety of services with explicit QoS requirements (the Internet traffic growing rate is of about 400% a year). Because of the finite amount of network resources, it becomes vital to support mechanisms that makes it possible to efficiently allocate and manage these resources in a way that QoS requirements can be met. This is concretely reinforced by all major router/switch vendors that provide specific QoS mechanisms in their products (see for instance [295], [317], [296]). Moreover, the Internet Engineering Task Force (IETF) has proposed many service models and protocols for providing QoS in the Internet, such as the Integrated Service and the Differentiated Service frameworks, the Multi-Protocol Label Switching (MPLS), the Resource Reservation Protocol (RSVP).

In addition, as argued by Frei in [93], from the routing point of view, the key resource to manage in networks is bandwidth for at least three reasons:

- Bandwidth is the primary QoS parameter for most (if not all) applications, and ensuring other QoS constraints, but not the throughput does not make sense.

- Bandwidth is a non-sharable resource: once allocated to a demand, the reserved bandwidth cannot be used by another demand.

- Most end-to-end parameters are usually not completely independent from each other because their behaviour depends on the throughput usage: if the load increases then delay, delay jitter or loss increase too.

1.1.3   Traditional Approaches to Multi-Provider Coordination

In the communication networks community, several bodies have already considered various problems in the area of the multi-provider network management and interactions. The need for standardised solutions resulted in the specification of the Telecommunication Management Network (TMN) architecture by the International Telecommunications Union [327] (ITU). The TMN architecture [140], [143] defines three main conceptual layers that address different needs for the organisation of a provider operations[1]. The *TMN information architecture* provides data representation of the network resources for the purpose of monitoring, control and management. The *TMN functional architecture* describes the realization of a TMN in terms of different categories of units and different classes of interconnection among these functional units. These classes are called *reference points* (RFs) and they can exist between different types of functional units and between functional units inside and outside of the same organisational domain. RFs provide the support for building interfaces between specific functional units at the *TMN physical architecture*, which corresponds to the physical realization of the TMN functional architecture. In particular, distinct network providers inter-operate by exchanging limited information concerning boundary network elements through the *TMN-X interface* [140], [144], see Figure 1.2. To ensure interoperability and end-to-end service management, the specification of this interface requires the use of identical communication protocols. The TMN-X interface deploys the Common Management Information Service [138] (CMIS) and the Common Management Information Protocol [139] (CMIP) that define management services exchanged between peer entities. By making use of CMIS primitives, CMIP provides a set of capabilities such as information filtering, asynchronous events report, network element configurations, etc. For any management system that exchanges information across a TMN-X interface, the scope of management activities must be well defined. It is the information model supported by each of the communicating TMN building blocks that defines the range of effects of management operations inside every distinct network domain. The information model used to share a common view of available management information and functions at the TMN-X boundary is also called Shared Management Knowledge [145]. Among various initiatives, the MISA ACTS AC080 Project [241] has developed an ATM/SDH network independent TMN-X interface for path-provisioning. One of the most significant contributions of this work is the definition of a unified framework for multi-domain network management that encompasses the heterogeneity of different underlying network technologies.

Another major contribution to the management of end-to-end connections in a multi-provider environment is given by the framework defined within the Telecommunications Information Networking Architecture (TINA) initiative [324]. TINA specifies an ubiquitous software platform for service logic, covering both

---

[1]More details about network management and the TMN framework can be found in Section 2.2.1.

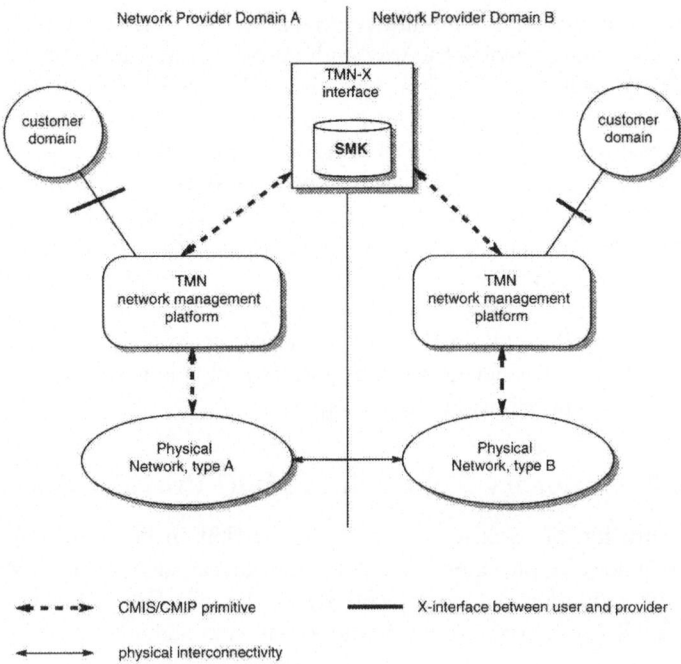

Figure 1.2: *In the TMN framework, interconnectivity and communication across hetero-geneous networks is achieved through the TMN-X interface. The set of possible services and operations is defined by using the available CMIS/CMIP primitives. The common information that two distinct provider domains share is called shared management knowledge (SMK).*

service operation (or service set up) and service delivery (or service manage-ment) [323], [243], [117]. In the TINA service architecture, the *Layer Network Coordinator* (LNC) is an object responsible for interconnecting termination points in a subnetwork defined as a domain. If several distinct domains need to intercon-nect to each other, there is an horizontal interaction between distinct LNCs [131]. Coordination is, in this case, achieved at the resource configuration management level. An alternative way to structure multi-domain connection management is the *federation* mechanism [227]. Instead of hierarchical relationships between compu-tational objects (as when deploying LNCs), federation defines peer-to-peer rela-tionships through the definition of standard interfaces. In the TINA context, this kind of interfaces are called Layer Network Federation (LNFed) inter-domain ref-erence points. Domain federation is considered a static contractual relationship between providers and represents a pre-requisite for establishing services spanning different domains. The REFORM AC208 project [255], for instance, has imple-mented a TINA compliant architecture for multi-domain connection management functionality.

Within the networking community, the TMF[2] [325] is currently making important contributions to evolve the way multi-provider interactions can take place. For this purpose, a group of several organisations is working on the integration and cohesion of Service and Network management operations within the so called SMART-TMN initiative [214]. The main goal of this initiative is to supply more flexible, dynamic and smart interfaces between customers and providers, between distinct providers, and between providers and network operators. A network operator is a service provider that owns and directly controls the deployed networks resources. A service provider may directly cover the role of network operator or subcontract it without taking care of managing the low level network resources such as links and nodes. The area that still raises several delicate open issues is the interface between distinct network operators and this is where this work aims to bring innovative contributions (see Chapter 8).

1.1.4   An Evolutionary Approach to Multi-Provider Coordination

The main limitation of traditional approaches is that in face of the growing and changing technology requirements low level interfaces, such as the TMN-X interface or the LNFed reference point, and static domain federation are becoming increasingly inadequate [68]. Network and service management heavily rely upon the direct control of human operators that are responsible for controlling inside their own domains low level network components through the usage of tools and/or application programming interfaces. At the inter-domain level, human controllers usually interact with each other via telephone, facsimile, e-mail, etc. This makes the overall interworking process very slow (for instance, several days can pass before effective inter-domain network configuration changes take place) and quite inefficient. Inefficiency is mainly due to the difficulty for human operators to consider all aspects which complicate the interworking process, such as:

- The increasing number of actors (content providers, added-value service provider, brokers, etc.) with different roles than those of the traditional Telecom operators. With market liberalisation the difficulties of pan-industry coordination clearly increase with the number of actors involved.

- The use of different and heterogeneous technologies and the technological rate of change. In today's networks, the number of technological options is continually growing. Consequently, the time between idea and implementation has been significantly reduced. There is often not enough time for the industry to develop guidelines before those guidelines are already obsolete.

- The need for an efficient mapping between intra- and inter-domain management aspects such as resource allocation, routing, pricing, etc.

---

[2]The TeleManagement Forum (TMF) is an industrial forum building on top of TMN standards.

Considering these aspects, what seems more suitable for future network scenarios is a management solution based on static and/or mobile software entities [343], [309], [194], [311] collecting network state information and which have the ability to directly invoke effective changes to switch controllers without the interaction of a human operator. Among various technologies, software agents have been argued to have a strong potential for better solving several problems in communication networks [167], [344], [123], since they can be autonomous, distributed, reactive, pro-active, intelligent, heterogeneous, self-learning and dynamic (see Section 8.4). In particular, [18], [235], [58], [33], and several contributions in [125] discuss the application of agent-based solutions for various tasks involved in the multi-provider service provisioning process such as connection configuration, routing, control, etc. Also the Foundation for Intelligent Physical Agents[3] (FIPA) envisages the *network management* scenario as one of the most interesting and significant agent application fields [87]. FIPA gives the basic guidelines about the fundamental functionalities that software entities should provide, but the details of how to effectively integrate different agents and improve the service provisioning process with intelligent techniques, i.e., coordination, negotiation, etc., are left to application developers.

Existing agent-based paradigms developed for routing multimedia traffic over distributed networks that are directly related to this work are examined in Section 2.3.2. In almost all these frameworks, negotiating agents acting on behalf of network providers (i.e., owners of the network resources) are used for defining network offers that are proposed to service providers (i.e., providers of networking services that do not necessarily own the underlying physical network resources) or directly to end user agents. However, how peer provider (neither network-to-network nor service-to-service) negotiations that can take place has not been exhaustively considered and several issues concerning automated coordination need more work [68]. As mentioned earlier, the coordination of peer providers is complicated by the need to formulate global end-to-end offers consistent with both inter- and intra-domain constraints without revealing strategic information to external agents. In traditional interworking approaches, negotiation between peer providers is usually based on a single offer at a time. The problem here is that the number of possible offers (i.e., intra-domain routes) grows combinatorially with the number of network providers involved in the service allocation. Therefore, making a good (or ideally optimal) end-to-end offer becomes a hard selection problem that is very difficult to solve with limited resources.

The NPI paradigm aims to address these main challenges by making possible:

- Dynamic inter-domain negotiations with the option of considering more than one route at a time without the need to reveal internal and confidential data. This is achieved by the usage of *Distributed Constraint Satisfaction Problem* (DCSP) techniques [368]. Constraint satisfaction is a powerful and exten-

---

[3]FIPA (www.fipa.org) is a non-profit standardisation group that aims to promote interoperability of emerging agent-based applications, services and equipments.

sively used Artificial Intelligence (AI) paradigm that involves finding values
for problem variables subject to restrictions (constraints) on which combinations of values are acceptable. Multi-provider service demand allocation
can be considered as a DCSP since the variables (local network resources
such as bandwidth, routers, etc.) are distributed among agents and since
constraints exist among them. More precisely, end-to-end routes are decomposed into fragments (i.e., distinct variables) corresponding to independent
decision makers. In DCSP terms, there is one variable per provider whose
values are route fragments through that provider. Constraints between the
variables ensure that route fragments connect and specific DCSP methods
are applied in order to rapidly access what choices can be part of a consistent end-to-end route. The space of possible solutions provides the basis for
subsequent negotiations. More than one local route at a time can be negotiated since a single variable can take as value one of a large set of end-to-end
routes. Decisions on the preferences of path fragments are made by finally
pruning out possible values from this set.

- Efficient intra-domain resource management and dynamic pricing. Within
  a single provider network, the possible routes are represented in a compact
  and flexible way by a Blocking Island (BI) structure [93]. The BI formalism
  provides a powerful resource abstraction technique that allows the quick assessment of the existence of routes between end-points with a given amount
  of available bandwidth, without having to explicitly search for such a route.
  Therefore, the BI structure can be used to speed up agent decision making
  by accelerating the process of determining the intra-domain solutions space,
  i.e., the set of possible local routes that can support the incoming service demand. Moreover, the BI formalism has been refined in order to dynamically
  estimate the opportunity cost of allocating specific network resources. Pricing services that span areas owned by different authorities, either Telecom
  networks or Internet domains, is a very complex task that has received an
  increasing attention over the last five years (see Section A.1). Ideally, prices
  should dynamically reflect the current network state and take into account
  the opportunity cost of deploying a specific amount of network resources.
  This is a very challenging issue especially in a highly dynamic environments
  where the network infrastructure should support an efficient routing method
  which is able to reflect the offered prices immediately. Current routing algorithms for large scale networks such as PNNI [91] and BGP [185] do not respond to this dynamicity need. Furthermore, another major difficulty is how
  to estimate the opportunity cost. This cost should dynamically reflect the
  probability of blocking future incoming demands when deploying a specific
  amount of network capacity, i.e., bandwidth on a link. In current communication networks, however, costs are mainly considered as constant amounts
  independently on the current resources availability. In Chapter 5, an innovative method for dynamically estimating costs that reflect the current resource
  utilisation state is proposed.

To summarise, this volume proposes an agent-based dynamic structure that efficiently integrates inter-domain coordination with intra-domain resource allocation and pricing. This structure is innovative in two main ways. First of all, the integration of inter- and intra-domain tasks is enforced by the use of distributed constraint satisfaction techniques combined with automated negotiations. Furthermore, the use of powerful clustering techniques simplify and improve network control, including resources allocation and pricing.

1.2 Beyond Automated Multi-Provider Coordination

The definition and the development of the NPI framework rely upon a combination of diverse methods and techniques from different fields. This combination is essential when considering the different needs that a suitable solution to automated multi-provider coordination implies. The main requirements that guided this work choice are *privacy* of strategic and internal information, *termination* of the algorithms and protocols developed, *efficiency* in terms of network performance and benefits for both providers and consumers and *stability* of the mechanisms developed (see Chapter 8). In the following, an overview of the main disciplines that contributed to the development of the NPI paradigm is given. A more complete background overview is given in Chapter 2.

1.2.1 Multi-Agent Systems and Coordination

Currently, agent technology represents one of the most dynamic fields in Computer Science and is likely to play a prominent role in almost all future computer systems [149], [30]. The main characteristics of the software agents developed within this work are summarised by the definition given in [355][4]. This definition suggests that an agent is:

> an encapsulated computer system that is situated in some environment
> and that is capable of flexible, autonomous action in that environment
> in order to meet its design objectives.

This means that agents are entities which receive inputs related to the state of the environment they are embedded in through sensors and they act on the environment through effectors. The main idea is that agents act without direct intervention from others (humans or other software processes), they exhibit opportunistic, goal-directed behaviour and take the initiative when appropriate. All input and output from an agent is considered as sensing and performing actions. Therefore, an agent does not directly receive commands from users, though users input influence its behaviour. Moreover, an agent requires the action to perform to be complementary to its goals. It can decide to perform or not tasks at specific conditions if not in contradiction with its own goals.

---

[4]Other popular formal definitions for agency can be found in [270], [193], [121], [356], [29].

The multi-provider service provisioning process can be naturally modelled as a multi-agent system (MAS) that is a system containing multiple agents that can interact and influence each other's behaviour [148]. Research in MAS is primarily concerned with the *coordination* of autonomous (possibly heterogeneous) and self-motivated computational agents.

In general, in the DAI framework, which includes both the cooperative distributed problem solving area and the MAS field, coordination has been examined by many researchers and several definitions have been proposed (for a survey of coordination in MAS Chapter 3 of [82] is recommended). While in cooperative distributed problem solving [25], [70], systems are centrally designed and interaction protocols and strategies are imposed for each agent, in MAS there is no global control, no globally consistent knowledge and no global goal [163], [266], [284]. Self-interested agents choose the best strategy for themselves, which cannot be explicitly imposed from outside. In this volume, coordination is directly concerned with mechanisms that regulate interactions of autonomous self-interested entities in the multi-provider network scenario. Among different types of coordination including organisational structuring, contracting, planning, etc. (see [340], [217] for more exhaustive surveys), *automated negotiation* is primarily considered.

1.2.2   Automated Negotiation

Many researchers from different fields have focused on the formalisation of automated negotiation. As a result, several definitions have been proposed [266], [177], [184]. In this volume, *negotiation* is considered as a process by which a joint decision is made by two or more parties which first verbalises contradictory demands and then moves toward agreement by a process of concession making or searching for new alternatives [244]. Typically, each party starts a negotiation process by offering the most preferred solution from an individual perspective. If the offer is not accepted by other parties, then a counter-offer can be made in order to converge towards an agreement. During this process, the set of available options for each agent decreases until an agreement is or is not reached.

When defining an *automated* negotiation system it is fundamental to choose the possible negotiation protocol(s) to design appropriate negotiation strategies. A negotiation protocol specifies the 'rules of encounter' between negotiation participants, which means the possible roles that participants can cover, what deals can be valid and the permissible sequences of actions, i.e., which messages are allowed and in which order. A negotiation strategy establishes the way an agent behaves in an interaction, i.e., what specific deals will be proposed during the negotiation. Given a pre-fixed protocol, there may exist several compatible strategies, each of which produces a very different outcome. A certain negotiation protocol together with the participants' strategies is also called *negotiation mechanism*.

Negotiation has been proved to be a powerful instrument for coordinating both self-motivated and/or cooperative agents. In this context, the focus is on interactions of self-interested entities, which have been extensively studied in microeconomics [330], [246], especially by the game theory community [247] [166], [99], and in the DAI field [73], [162], [69], [266]. Most of these approaches, also called the *game theoretic models*, assume perfect rationality of 'players', which means that agents are considered capable of flawless deduction, optimal reasoning about future circumstances and recursive modelling of other agents beliefs. In addition, in many cases, perfect computational rationality is a fundamental condition. This means that no computation is required to find mutually acceptable solutions within a feasible range of outcomes. More recently, increasing attention has been given to models where a cost is associated with both computation and decision making. In these approaches, agents are considered as rational bounded entities, which means that computational and cognitive limitations are taken into account. Examples of microeconomic models of bounded rationality can be found in [110], [305], [130], [104], [269].

Among various existing bounded rational approaches for automated negotiation a further distinction can be made:

- *Heuristic Models*: a cost is associated with both computation and decision making. In particular, every agent decision making process is modelled heuristically during the course of negotiation: the chosen protocol does not prescribe an optimal course of action and the aim is to produce good solutions rather than optimal ones. Relevant work in this area is presented in [13], [85], [165], [286], [319].

- *Normative Models*: again a cost is associated with both computation and decision making, but here the aim is to establish the best strategy that an agent can use within a given interaction protocol. In these approaches, negotiation mechanisms can be proved to have a certain number of properties such as stability, Pareto efficiency, etc. Interesting explicit normative models of bounded rationality include [172], [279], [173]. For an overview, Rubinstein's book [269] on modelling bounded rationality is suggested.

The negotiation framework developed within this context is a heuristic model that falls into a particular category of bounded rationality where boundaries, limitations and constraints are intrinsic in the CSP-based decision making model that the developed agents rely upon. In addition, in the NPI paradigm agents have limited computational resources (i.e., a cost is associated to both computation and communication efforts) and finite time to converge to an agreement (i.e., fixed negotiation timeouts). Two main alternative strategies relying upon an innovative way of estimating resource criticalness in the network have been proposed (as discussed in Chapter 5) and experimentally validated (see Chapter 7). Furthermore, the combination of the NPI coordination paradigm with the CSP-based decision making model has been shown (in Chapter 8) to be able to lead to a Nash solution [209].

1.2.3   Distributed Constraint Satisfaction Techniques

Constraint satisfaction is a powerful AI paradigm which proposes techniques to find assignments for problem variables subject to constraints on which only certain combinations of values are acceptable. The success and the increasing application of this paradigm in various domains mainly derive by the fact that many combinatorial problems can be expressed in a natural way as a *Constraint Satisfaction Problem* (CSP), and can subsequently be solved by applying powerful CSP techniques [338]. Resource allocation, scheduling, planning, configuration and diagnosis are typical examples of problems that have been modelled and in many cases efficiently solved with CSP.

A finite discrete CSP is defined by a finite set of *variables* whose values are taken from finite, discrete *domains* and a set of *constraints* restricting the values that variables can simultaneously take. *Unary* constraints restrict the domain of a variable without reference to any other variable. For instance, the condition "$v > 3$" expresses a unary constraint for variable v. *Binary* constraints restrict the values a variable v_1 can take by comparing it to another variable v_2. For instance, "$v_1 < v_2$" is a binary constraint between variables v_1 and v_2. An *arc* represents an existing binary constraint between two specific variables. The *arity* of a CSP is defined by the maximal number of variables involved in a constraint (more formal definitions can be found in [192] and [65]). This basic CSP definition has previously been extended in many ways allowing dynamic sets of variables, dynamic, continuous or infinite variable domains and constraints of various arity. A simple and intuitive example of CSP is given by the map colouring problem. In this problem, each region of a map needs to be coloured by selecting a specific colour from a given finite set, e.g., C = {white, grey, black}, so that adjacent regions have different colours. In the equivalent CSP formulation, there is a variable for each region of the map. The domain of each variable is the given set of colours C. For each pair of regions that are adjacent in the map, there is a binary constraint between the corresponding variables that disallow the same assignment of colour to the two adjacent regions.

Solving a CSP means finding an assignment of values to all variables in a way that none of the constraints are violated. The techniques for solving CSPs can be subdivided in two main groups: search (e.g., *backtracking* and *iterative*) algorithms and inference (e.g., *consistency*) methods [191]. Consistency algorithms are pre-processing procedures that are invoked before search algorithms. Backtracking methods construct a partial solution (i.e., they assign values to a subset of variables) that satisfies all of the constraints within the subset. This partial solution is expanded by adding new variables one by one. When for one variable, no value satisfies the constraints between the partial solution, the value of the most recently added variable is changed, i.e., backtracked. Iterative methods do not construct partial solutions. In this case, a whole flawed solution is revised by a hill-climbing search. States that can violate some constraints, but in which the number of constraint violations cannot be decreased by changing any single variable value (i.e., local-minima) can be escaped by changing the weight of constraints and/or restarting from another initial state. Iterative improvement is efficient but not complete.

A distributed CSP is a CSP in which the variables and constraints are distributed among distinct autonomous agents [368]. Each agent has one or multiple variables and tries to determine its/their value/s. In general, there exist intra- and inter-agent constraints and the value assignment must satisfy all these constraints. In order to verify inter-agent constraints, i.e., constraints between variables controlled by distinct agents, some form of communication needs to be supported. The most known and deployed communication model which the DCSP paradigm relies upon assumes that agents communicate by sending messages, the delay in delivering messages is finite, though random, and messages are received in the order in which they are sent [368]. Many DAI problems that require finding a consistent combination of agent actions can be formalised as DCSPs. Among various works, some significant examples include: a multi-agent truth maintenance system [134], a distributed interpretation problem framework [199], a distributed job-shop scheduling [310] and a nurse time-tabling task [183] systems, a resource allocation approach in distributed communication network [57], [213], [372] and frequency assignment for cellular mobile systems [371]. Surveys of the most well known algorithms for DCSP are [374] and [369]. The most trivial algorithms for solving DCSPs are the *centralised method* and the *synchronous backtracking*. In the first case a leader agent gathers all information about the problems, their domains and their constraints and then solves the CSP alone using classic centralised constraint satisfaction algorithms. This approach is very inefficient since there are costs associated with gathering information, there are unused resources (i.e., agents sitting idle) and there is no confidentiality of data including both constraints and variable domains. The synchronous algorithm assumes that agents agree on an instantiation order. The first agent generates a partial solution that is submitted to the second agent, which generates an extension to the partial received solution and sends it to the third agent and so on. If the solution cannot be extended then a backtracking message is sent to the previous agent. With this approach there are still costs to decide the transmission of the instantiation order; furthermore, only one agent is active at a time. In order to overcome these major limitations, several *asynchronous search* algorithms that allow agents to work in parallel and asynchronously have been developed [368], [372], [373], [118].

In this volume, multi-provider service demand allocation has been modelled as a DCSP since the variables (local network resources such as bandwidth, routers, etc.) are distributed among agents and since constraints exist among them. More precisely, end-to-end routes are decomposed into fragments (i.e., distinct variables) corresponding to independent decision makers, see the example depicted in Figure 1.3. In DCSP terms, there is one variable per provider whose values are route fragments through that provider. Constraints between the variables ensure that route fragments connect and specific DCSP methods are deployed in order to rapidly access what choices can be part of a consistent end-to-end route. The

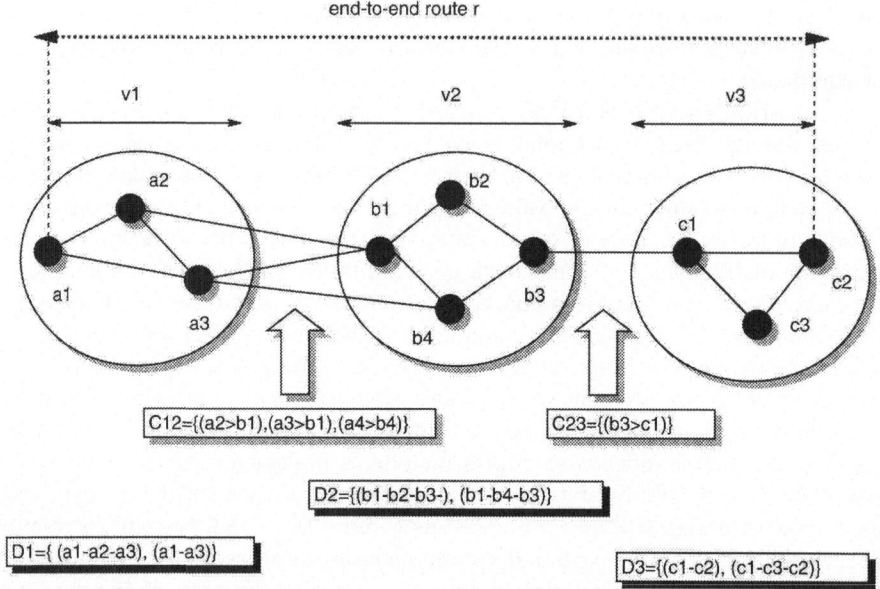

Figure 1.3: *The end-to-end route r is decomposed in three main fragments that correspond to the three distinct variables v1, v2 and v3. The variable domains D1, D2 and D3 collect all possible values for v1, v2 and v3 respectively. Constraints between adjacent network domains are expressed as connectivity requirements between boundary network nodes (set C12 and C23).*

space of possible solutions provides the basis for subsequent negotiations. More than one local route at a time can be negotiated since a single variable can take as values a large set of end-to-end routes. Decisions on the preferences of path fragments are made by finally pruning out possible values from this set.

To summarise, NPI agent decision making is modelled as constraint-based reasoning and the coordination of self-interested agents is enforced by constraint propagation methods that maintain the range of options (possible deals) considered by each agent during the negotiation process.

1.2.4   Abstraction Techniques

Abstraction methods and problem reformulation have been extensively proposed as promising techniques to reduce the complexity of problem solving and knowledge representation. The main idea is to build abstractions that effectively structure the problem space and map the current problem representation into a simpler one so that reasoning in this *abstract space* is less complex than in the original one (see [109] and [348] for formal definitions of abstraction).

Abstraction and reformulation methods have been applied to a large number of domains[5] such as game reformulation [210], theorem proving [232], [109], problem solving [328], logic programming [60], planning [157], [160], casual simulation [347], digital circuit troubleshooting [119], communication models [75], data bases [307], [180], cooperative query answering [48], document processing and management [200], and model-based diagnosis [206].

These kinds of techniques are also extensively used in communications networks. Several works have discussed the importance and the effectiveness of resources abstractions for reducing the complexity of routing, control and management tasks (see, among others, [311], [178], [11]). Additionally, in many deployed networks such as the Internet and ATM networks, routing protocols heavily rely upon the usage of specific information summaries, or aggregated views, of the costs and availabilities of traversing network elements and domains [8], [11]. In [46], abstraction methods support reactive strategies for the management of network resources. The aim is to improve the performance of typical network management and control functions by identifying bottlenecks and conflicts. Allemand and Liver [2] propose building decision support systems in the telecommunication networks area by deploying abstraction techniques. Abstractions between operators and machines (i.e., computers) and between machines and end-users are captured and organised as the basis for decision support systems.

Within the activity carried on in this context, the most relevant work concerning abstraction techniques for communication networks is the approach discussed by Frei in [95]. Frei defined a clustering scheme based on Blocking Islands (BIs), which can be used to represent bandwidth availability at different levels of abstraction, as a basis for resource allocation in communication networks. In particular, BIs-based techniques can be used to effectively solve off-line QoS routing problems [93]. This approach works by dynamically reformulating an aggregated view of the network state that is the basis of the solving process. How to combine this off-line, centralised technique with coordination methods to create a distributed use of abstractions capable of solving on-line routing problems inside every communication network domain has been addressed by Willmott [353]. In this volume, BIs-based techniques are used to quickly assess inside every network the existence of routes between end-points with a given amount of available bandwidth, without having to explicitly search for such routes. This makes it possible to speed up the coordination process between distinct operators by pruning out all inconsistent solutions from the set of intra-domain routes considered during negotiations with peer providers. Furthermore, in this context, the BIs paradigm has been re-elaborated for defining the *Availability Criticalness Evaluation* (ACE) paradigm, which enables the definition of a *criticalness* factor dynamically taking into account the probability of blocking future incoming demands by deploying re-

---

[5]Many applications of abstractions methods are reviewed in [109] and [47].

sources on specific links. This probability corresponds to the risk of isolating some network nodes in terms of reachability in a BI-based network resources structure called *Blocking Island Hierarchy* (BIH). By deploying the BIH structure, the ACE mechanism becomes able to map network criticalness into prices and pricing becomes an instrument to effectively control network resource allocation (see Chapter 5).

1.3   Volume Outline

The rest of the book is organised as follows.

- Chapter 2 reviews background principles of communication networks, introduces multi-provider scenarios and discusses related work in the fields relevant to this framework. This includes an overview of the major existing interworking frameworks; a survey of the significant works that deploy software agents for automated control and management in communication networks; a description of negotiation-based methods for modelling and solving coordination of individual self-interested entities; and, to conclude, a review of the basic principles and techniques in the DCSP field.

- In Chapter 3, a preliminary step to avoid ambiguity is the definition of several fundamental terms and concepts used within the whole book. This enables a formal definition of the multi-provider service set up problem, including a model of the underlying communication networks, the types of services supplied, and the end-to-end communication demands. The final part of this chapter concentrates on the formalisation of the intra-domain tasks that every provider has to perform when allocating network resources. The focus is on the clustering scheme that has been deployed to represent bandwidth availability at different levels of abstraction. This enables the reduction of the complexity of resource allocation and pricing tasks.

- The description of the NPI agent-based framework is given in Chapter 4. The first part introduces the conceptual model behind the NPI approach, including the definition of the different types of agents developed, their roles, their objectives and their internal architecture. The second part of the NPI description discusses the main agent interactions that enable coordination of self-interested entities to be achieved in different phases of the service setup process. This leads to description of the distributed solving process upon which the network-to-network operators coordination paradigm relies. This includes the presentation of the specific DCSP-based techniques deployed for rapidly accessing what choices (i.e., local routes) can be part of consistent end-to-end routes spanning multiple networks.

- Chapter 5 discusses the integration of economic principles within the various self-interested software entities acting on behalf of users and providers. This includes the presentation of interaction protocols used, a description of the decision making models that the NPI agents rely upon and the definition of the strategies they employ. The analysis of the providers' negotiation behaviour leads to the second part of the chapter in which the main issues of pricing communication services are examined. As network technologies mature, price tends to become one of the dominant selection factors for purchasers of network services. To compete on prices, providers must reduce the cost of provisioning and maintaining these services. In this framework, the *Availability Criticalness Evaluation* pricing policy is proposed as an innovative way of evaluating costs and defining prices.

- Alternative approaches for the coordination of distinct providers are described and discussed in Chapter 6. The main objective is to underline weaknesses and strengths of different solutions with respect to the NPI approach. There are two main alternative solutions that are considered in this context. The *Fixed Agreements Solution* (FAS) is a decentralised approach in which peer-to-peer coordination is regulated by pre-existing static agreements and where provider do not dynamically negotiate. The *COalition Based* (COB) approach relies upon the dynamic coordination of providers forming on-demand coalitions supervised by a central controller.

- Chapter 7 presents the empirical results obtained to evaluate the performance of the NPI techniques including benchmarks with alternative solutions. The first part introduces the test scenarios, including the definition of specific performance descriptors. The second part presents two main sets of experimental results. The *networking set* estimates typical networking performance. The *beneficial set* analyses the benefit of the various entities involved in the service provisioning from a market perspective.

- These results are qualitatively and quantitatively discussed by considering several criteria that are proposed in Chapter 8. In this chapter, the analysis of the NPI paradigm and the discussion of the results obtained through experiments make it possible to underline its main strengths. In particular, various interesting aspects emerge from the discussion of the relative performance of the alternative coordination approaches tested. This critical and comparative analysis of the NPI paradigm also enables the extrapolation of the main open issues of this work. The focus here is on the aspects that, despite the relatively exhaustive results confirming the value of the NPI techniques, still require more investigation.

- Chapter 9 concludes this book by discussing the main contributions of this work, summarising the main limitations of our solution and finally presenting future research directions.

- Appendix A reviews background principle of pricing resources in communication networks and gives a short summary of relevant related works.

- Finally, Appendix B lists the acronyms and abbreviations used in this volume.

Chapter 2

Background

*A problem faced in most reasoning situations is that
all the information that may be relevant is not available and that
what is available is confusing and not necessarily relevant.*
– RAJ BHATNAGAR AND LAVEEN N. KANAL, *1992*

This Chapter reviews background principles of communication networks and discusses the most relevant research behind this work. Fundamental concepts are recalled by introducing the most significant network infrastructures and technologies existing today (Section 2.1). Particular attention is payed to multi-provider environments where usual networking operations are complicated by the need for coordinated interactions with external domains (see Section 2.2). An overview of the major existing multi-domain management frameworks makes it possible to characterise which specific intra- and inter-domain tasks represent the major challenges in current scenarios. This leads to an analysis of several interworking solutions that have been developed in the last five years and that make use of software agents to automate specific management or control operations (see Section 2.3).

In the second part of the Chapter, significant research that has been undertaken in fields which have significantly contributed to the definition of the NPI approach is considered. More precisely, this includes an overview of relevant work in the following two areas:

- *Multi-agent systems and coordination*: The focus is on negotiation-based methods and for modelling and solving coordination of individual self-interested entities. In particular, previous significant works that deploy software agents for automated interworking in communication networks are reviewed (see Section 2.4).

- *Distributed constraint satisfaction techniques*: After a short review of basic principles that allows the definition of a Distributed Constraint Satisfaction Problem (DCSP), the most relevant works that deployed DCSP based techniques combined with negotiation mechanisms are discussed (see Section 2.5).

2.1   Communications Networks

A communication network is the combination of numerous network elements that are required to support voice, data, or video services in local or long-distance applications. Networks can vary both in size and complexity ranging from Local Area Networks (LANs) and Wide Area Networks (WANs), through to Internets (a network of networks) and Intranets (a private Internet). LANs and some Intranets exist in one physical location, whereas WANs and Internets are much larger, both in power and geographic size. In a communication network two basic components are *terminals* and *servers*. A terminal is a device (e.g., phone, computer) that is connected to a network. A server is a device that has software that allows it to connect terminals together and share hardware devices (e.g., router, gateway). A server controls the network and it may also have public files on it (in which case it may be called a file server).

A network bearer services comprise the end-to-end transport of traffic, in specific formats, over a set of routes. The fundamental mechanisms that enable these services are: *multiplexing, switching* and *resource allocation*.

Multiplexing is the process of sending multiple signals or streams of information on a carrier at the same time in the form of a single, complex signal and then recovering the separate signals at the receiving end. Analog signals are commonly multiplexed using frequency-division multiplexing (FDM), in which the carrier bandwidth is divided into sub-channels of different frequency widths, each carrying a signal at the same time in parallel. Digital signals are commonly multiplexed using time-division multiplexing (TDM), in which the multiple signals are carried over the same channel in alternating time slots. In some optical fiber networks, multiple signals are carried together as separate wavelengths of light in a multiplexed signal using dense wavelength division multiplexing(DWDM).

Switching is the combination of hardware and software processes that move data coming in to a network node out by the correct port (door) to the next node in the network. In the following, the two most important switching techniques, circuit switching and packet switching, are explained in more detail.

Resource allocation is the process by which the limited amount of network resources, such as link bandwidth and switch buffers, ensures that each application receives the necessary resources to maintain its quality of service.

In the networking community, one of the most common classifications distinguishes between two types of networks that mainly differ in how data is transferred between distinct nodes: *circuit switched* networks and *packet switched* networks.

- In *circuit switched* networks a dedicated circuit (or path) is established between any two nodes of the network willing to communicate to each other. For the duration of the connection, all resources on that circuit are unavailable for other users. The traditional public switched telephone network (PSTN) as well as the first generation of Integrated Services Digital Network (narrow ISDN) are the most significant examples of this type of network.

- In *packet switched networks* data traffic is segmented into small units of data called packets (or datagrams or cells) that are routed through a network based on the destination address contained within each packet. Breaking communication down into packets allows the same data path to be shared among many users in the network. The main distinction is between connection-oriented and connection-less networks.

  - In *connection-oriented* infrastructures a logical connection is established before transferring all the packets along the established connection, see Figure 2.1. Since all data packets traverse the same sequence of network nodes (i.e, gateways), the packets arrive in order. Typical examples of this kind of networks are TDM, SONET/SDH, X.25, Frame Relay and ATM.

  - In *connection-less networks* packets belonging to one connection do not necessarily traverse the same sequence of network nodes (or gateways), see Figure 2.2. Datagrams from node 1 to node 2, for instance, can take different routes and a routing decision is made separately for each packet, possibly depending on the traffic at the moment the packet is sent. The Internet is the most known example of connection-less packet switched network using the Internet Protocol (IP).

2.1.1   Connection-oriented Networks

Connection oriented networks can be further classified by considering the specific technology that is used to either transport, manage and/or deliver data traffic. An exhaustive overview of these technologies is however out of the scope of this volume. Therefore, only a brief summary that aims to introduce fundamental networking concepts and some possible scenarios where our techniques could eventually apply is given. Relevant references are also given for those readers interested in more detailed information.

TDM [156] is a scheme in which numerous signals are combined for transmission on a single communications line or channel. Each signal is broken up into many segments, each having a very short duration. The circuit that combines signals at the source (transmitting) end of a communications link is known as a multiplexer. It accepts the input from each individual end user, breaks each signal into segments, and assigns the segments to the composite signal in a rotating, repeating sequence. The composite signal thus contains data from all the end users.

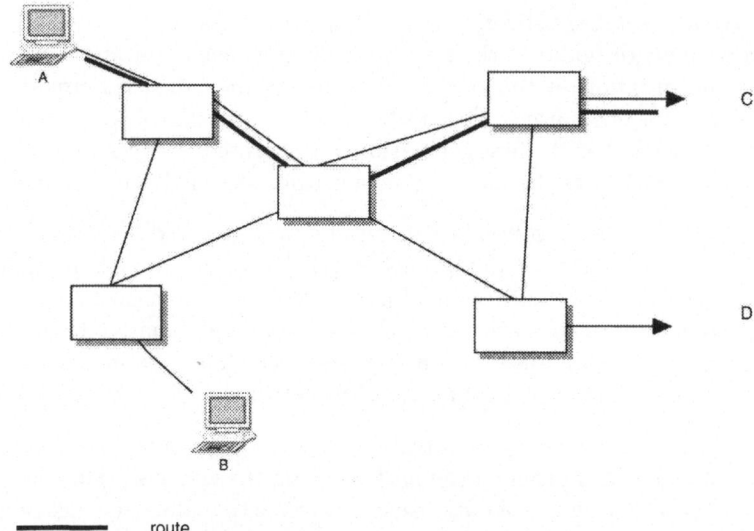

Figure 2.1: *In order to transmit information, a circuit switched network finds a route along which enough resources can be allocated and dedicated to the transmission of that specific amount of information. The network connects circuits together and reserves them for the transmission.*

At the other end of the long-distance cable, the individual signals are separated out by means of a circuit called a demultiplexer, and routed to the proper end users. A two-way communications circuit requires a multiplexer/demultiplexer at each end of the long-distance, high-bandwidth cable.

SONET [15] is the U.S. American National Standards Institute (ANSI) standard for synchronous data transmission on optical media. It encodes bit streams into optical signals that are propagated over optical fiber. The high speed of SONET and its frame structure make it possible to support a very flexible set of services. The standard specifies the frame structure as well as the characteristics of the optical signal. The most important feature of this standard is that all clocks in the network are locked to a common master clock, so that the simple TDM scheme can be used. The international equivalent of SONET is the synchronous digital hierarchy **SDH** [100] framework. Together, they ensure standards so that digital networks can interconnect and inter-operate internationally and existing conventional transmission systems can take advantage of optical media through tributary attachments [300].

The **X.25** protocol [358], adopted as a standard by the Consultative Committee for International Telegraph and Telephone (CCITT), is a packet-switching technology which was designed for transmitting analog data such as voice conversations. The protocols proposed by X.25 correspond closely to the data-link and physical-layer protocols defined in the Open Systems Interconnection (OSI)

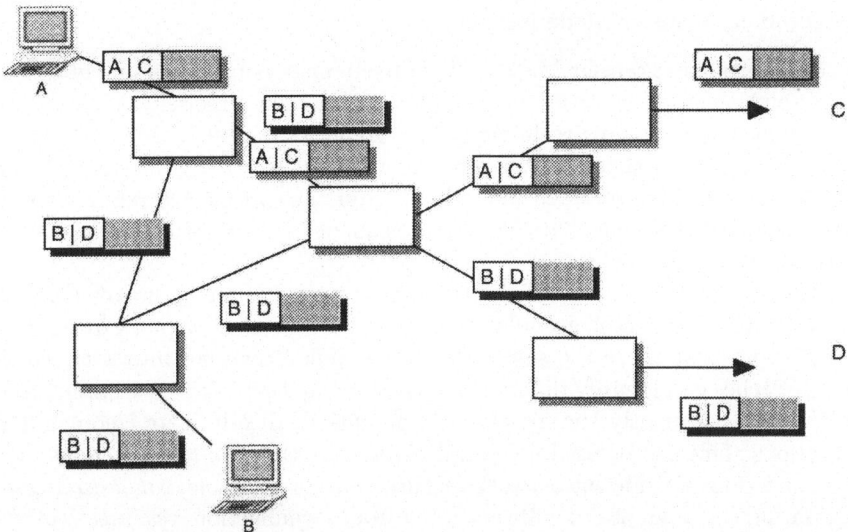

Figure 2.2: *In connection-less networks, packets are transported individually, potentially along different routes. Each packet contains its source and destination address so that intermediate nodes can properly forward it.*

communication model [359]. The *data-link* layer assures that an initial connection has been set up, divides output data into data frames, and handles the acknowledgements from a receiver that the data arrived successfully. It also ensures that incoming data has been received successfully by analysing bit patterns at special places in the frames. The *physical layer* supports the electrical or mechanical interface to the physical medium. For example, this layer determines how to put a stream of bits from the upper (data link) layer on to the pins for a parallel printer interface, an optical fiber transmitter, or a radio carrier.

Frame Relay [42], which is based on the older X.25, is a a fast packet technology with a committed information rate (CIR). Data is put in a variable-size unit called a *frame* and any necessary error correction (retransmission of data) is left up to the end-points, which speeds up overall data transmission. When an error is detected in a frame, it is simply dropped. The end points are responsible for detecting and retransmitting dropped frames. For most services, the network provides a permanent virtual circuit, which means that the customer sees a continuous, dedicated connection without having to pay for a full-time leased line, while the service provider figures out the route each frame travels to its destination and can charge based on usage.

Asynchronous Transfer Mode Networks

The Asynchronous Transfer Mode (ATM) technology is probably one of the most known solutions for connection oriented networks. More precisely, ATM is a dedicated-connection switching technology that organises digital data into 53-byte cell units and transmits them over a physical medium using digital signal technology [63]. Each cell is processed individually and asynchronously relative to other cells belonging to the same data block and is queued before being multiplexed over a specific transmission path.

Transmission paths in ATM networks are structured in two distinct hierarchical levels. The finest granularity connection is called *virtual circuit* (VC)[1]. Multiple VCs that share the same path are bundled together into a *virtual path* (VP). Switches can identify different connections by their Virtual Channel Identifier (VCI). Consequently, the switches can potentially discriminate among different connections. This can be used in several ways: *admission control*, i.e., refusing certain connections if sufficient network resources are unavailable, *congestion control*, i.e., limiting the amount of traffic accepted from a connection, *resource allocation*, i.e., negotiating the bandwidth and buffers allocated to a connection, and *policing*, i.e., monitoring the 'burstiness' and average rate of traffic in a connection.

The official routing algorithm for ATM networks is the Private Network-to-Network Interface (PNNI) [8]. PNNI specifies a topology state (link-state) algorithm in which nodes flood QoS characteristics and reachability information so that all nodes obtain knowledge about the state of the network and available network resources. To reduce the complexity, PNNI deploys a hierarchical model for topology aggregation [8]. Switching systems are organised into logical collections called peer groups. Every peer group has a peer group leader that represents the group in the parent peer group (at a higher level in the PNNI hierarchy) as a single node. The aggregation of information in this hierarchy allows route computation and information storage to be shared between network nodes, which improves the scalability of this kind of approach. Some important performance metrics used by this routing protocol are call setup delay, call blocking probability and resource utilisation efficiency. Call establishment in PNNI consists of two fundamental processes: the selection of an end-to-end path and the setup of the connection state at every network node along the path.

The ATM Forum specifies different categories of services that an ATM network can provide:

- Constant Bit Rate (CBR) services require information to be transferred between source and destination at a constant bit rate after virtual connection has been set up (e.g., digital voice and digital video).

- Variable Bit Rate (VBR) services data can be sent at an average sustained cell rate with burst allowed up to a peak cell rate.

---

[1]Remark that a connection over a virtual circuit is called a *virtual channel* in ATM terminology.

- Available Bit Rate (ABR) services are assured a minimum cell rate, but also provide feedback by the network if it is possible to send cells at a higher rate. This class is recommended for transfer of data which is sensitive to loss, but not to delay.

- Unspecified Bit rate (UBR) services provide a best-effort delivery with no guarantees about QoS.

Support for QoS requirements is guaranteed for CBR, VBR and ABR traffic. Connections can usually ask for bounds on performance metrics, such as cell loss rates, cell transfer delay, cell delay variation, etc.

Although, ATM was conceived to be the unifying protocol that would provide support for all types of traffic over an integrated network infrastructure, nowadays, its most dominant use involves providing connectivity for IP networks. The vast majority (roughly 80 percent) of the world carriers use ATM in the core (backbone) of their networks. At this level, ATM has been widely adopted because of its flexibility in supporting a broad array of diverse technologies, including DSL, IP Ethernet, Frame Relay, SONET/SDH and wireless platforms.

For more details about ATM standards, networks and technology we suggest the ATM Forum and the International Telecommunications Union Web sites [90], [327].

2.1.2 Connection-less Networks

In connection-less networks, when traffic has to be sent between two distinct network nodes, the full destination address and other control information required by the network are attached, as a header, to the message data to form a variable-length packet, or datagram. No procedure is initiated to establish a connection between the two users. Packets are launched into the network and the header address is used directly for routing them until the proper destination is reached. Therefore, in connection-less networks, connectivity is guaranteed by routing and forwarding datagrams rather than setting up and releasing connections like in connection-oriented environments.

The Internet

The Internet is basically a network of networks, connected by everything from modems and ordinary telephone lines to fiber optic cables, microwave links, and satellite transmissions. Important network nodes are hubs, bridges, routers and gateways. *Hubs* link groups of computers somewhat like a LAN, *bridges* connect LANs together. A *router* is more complicated than a bridge in that it can make decisions about where and how to send packets of information. A *gateway* is a network device that connects LANs that may be running on different network operating systems and that therefore might have to translate data between networks of different types. Usually, a gateway connects to a high-speed network cable or medium called a *backbone*, i.e., gateways act as interfaces between small networks and much larger ones, such as a LAN connecting to the Internet. The main difference between routers and gateways is that the former work at the network level and the latter at the application level. This means that routers pass messages from on network or media to another without any translation of the content, while gateways translate data between networks of different types.

Traffic routing in the Internet is organised on the basis of distinct domains called autonomous systems (ASs), which are identified by a globally unique number, i.e., the Autonomous System Number (ASN). Networks within an autonomous system communicate routing information to each other using intra-domain routing protocols, or Interior Gateway Protocols (IGPs), such as OSPF [205], IGRP [127], IS-IS [37] and RIP [126], while distinct autonomous systems share routing information using a border gateway protocol (see Figure 2.3). An autonomous system is also referred to as a single network or a group of networks that are controlled by a common network administrator (or group of administrators) on behalf of single administrative entity (such as a university, a business enterprise, or a business division).

The Internet was primarily targeted for data transport. Only later, it was used to transport other services such as media streams over the basic best-effort service. Although, this lead to extensive work in application adaptiveness, recently several enhancements to the Internet service architecture have been proposed. In particular, two major frameworks for supporting different classes of services within the Internet are the Integrated Service [27] and the Differentiated Service [23] models. These approaches basically provide users with new services that range from 'slightly better than best-effort' to strictly guaranteed QoS.

Internet Protocol IP establishes how data is sent from one network node to another on the Internet. As anticipated earlier, IP is a connection-less protocol, which means that there is no continuing connection (i.e., circuit) between the end points that are communicating. Each packet that travels through the Internet is treated as an independent unit of data without any relation to any other unit of data. Each computer (also called *host*) on the Internet has at least one IP address that uniquely identifies it from all other computers on the Internet. When data traffic is sent or received (for example, an e-mail or a file transfer), the message gets divided into packets. Each of these packets contains both the sender Internet address and the receiver address. Any packet is sent first to a gateway that understands and knows only a small part of the Internet structure. This gateway reads the destination address and forwards the packet to an adjacent gateway that in turn reads the destination address and so forth across the Internet until one gateway recognises the packet as addressed to a computer within its immediate neighbourhood or domain. That gateway then forwards the packet directly to the computer whose address is specified. Because a message is divided into a number of packets and each packet can, if necessary, be forwarded along a different route,

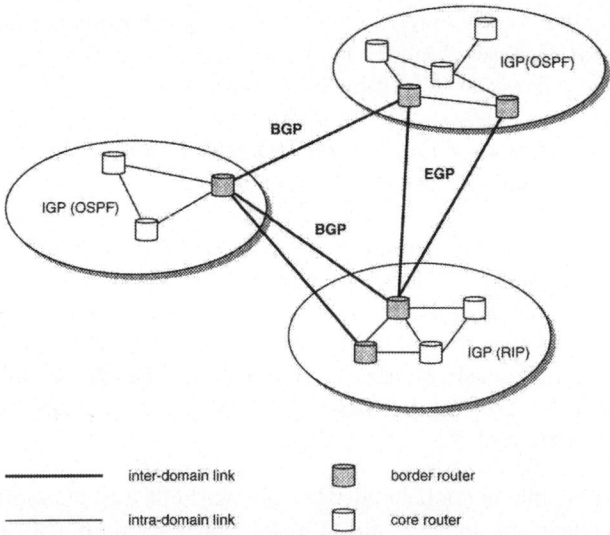

Figure 2.3: *Routing between distinct autonomous systems is performed through a border or exterior gateway protocols, i.e., BGP or EGP. Within a specific autonomous system and Interior Gateway Protocol (IGP) is used.*

packets can arrive at the destination in a different order than the order they were sent in. The Internet Protocol just delivers them. It is up to another protocol, the Transmission Control Protocol (TCP) [236] to put them back in the right order.

Today, the most widely used version of IP is the Internet Protocol version 4 (IPv4). However, the more recent IP version 6 (IPv6) [313] is also increasingly used and supported. IPv6 provides for much longer addresses and therefore for the possibility of many more Internet users. IPv6 includes the capabilities of IPv4 and any server that can support IPv6 packets can also support IPv4 packets.

The Integrated Service Model The *Internet Integrated Service* (IIS) framework (see the IETF specifications) [27] is an architecture that aims to enhance Internet services with network resource reservations for micro flows. In the IIS framework, a data flow is a sequence of messages that have the same source and destination (one or more) nodes and QoS requirements. These requirements are communicated through the network via a flow specification, which is a data structure used by router and/or gateways to request special services from the network infrastructure. A flow specification often guarantees how the traffic will be treated in the network (i.e., at specific nodes in the network). The main difference with connection-oriented systems, both circuit switched and/or packet switched, is that the IIS approach uses the same network and transport layer as the best-effort Internet. ISS is based upon the explicit reservation of network resources on hosts and routers in order to guarantee a defined set of QoS parameter values. Reservations are enforced on network nodes by deploying packet queueing and scheduling mechanisms that allow the allocation of a specific amount of network resources (e.g., bandwidth) to a single or a set of micro flows [64]. More precisely, the IIS approach deploys the *Resource Reservation Protocol* (RSVP), in order to implement resource reservation and the associated signalling [242].

The RSVP protocol [28] is a network-control protocol that enables Internet applications to obtain some QoS guarantees for their data flows. It is important to make it clear that RSVP is not a routing mechanism; instead, it works in conjunction with routing protocols and installs the equivalent of dynamic access lists along the routes that routing protocols calculate. RSVP supports three traffic types: best-effort, rate-sensitive, and delay-sensitive. The type of data-flow service used to support these traffic types depends on which specific QoS parameters are controlled and/or guaranteed.

- *Best-effort* traffic is traditional IP traffic without any QoS guarantee. Best-effort applications include file transfer, such as mail transmissions, disk mounts, interactive logins, and transaction traffic. The service supporting best-effort traffic is called *best-effort service*.

- *Rate-sensitive* traffic requires guaranteed rate and does not have stringent timeliness needs. An example of rate-sensitive application is H.323 video-conferencing. H.323 encoding [326] is a type of service that requires a constant transport rate. The RSVP service supporting rate-sensitive traffic is called *guaranteed bit-rate service*.

- *Delay-sensitive* traffic is characterised by the stringent timeliness of delivery requirements and it varies in its rate accordingly, e.g., MPEG-II video. RSVP services supporting delay-sensitive traffic are referred to as *controlled-delay service* (non-real time service) and *predictive service* (real-time service).

The fundamental limit of the IIS approach is in its scalability since the admission control for every micro flow is performed hop-by-hop and the amount of state information to be maintained at every hop increases proportionally with the number of flows. In addition, every router must support RSVP, admission control, multi-field classification and packet scheduling. Therefore, RSVP incurs a significant processing overhead and raises strong scalability concerns for large networks. The problem is amplified when demands cross domains controlled by different providers: there is no incentive for a provider to give priority to micro flows of another provider customer and there is no way to dynamically negotiate flows priority between peer operators.

The Differentiated Service Model The *Differentiated Service* (DS) framework [297] has been proposed by the IETF community in order to overcome scalability limits of the IIS approach and provide a globally scalable service architecture, based on an Internet made up of independently administered domains.

The principal idea behind the overall DS approach is to push the complex processing and resource management to the edges while keeping packet forwarding in core networks as simple as possible. For this purpose, there is no connection state in core (intra-domain) network nodes, but forwarding behaviour is only based on packet markings identifying the quality class. A specific byte in every IP-packet is used to mark data according to a customer profile in order to discriminate among different services. Such a profile is used to define which amount of packets belong to what service class and level. Unlike per-flow scheduling of packets, like in the mechanism proposed by the IIS framework, DS provides different service levels for traffic classes. This allows the aggregation of flows, which is fundamental to scalability. When there is no congestion, all packets are forwarded with the same priority.

In the DS framework, there are three different possible services.

- The *best-effort* service is the currently used approach in the Internet. There is no guarantee for QoS and everybody gets the service that the network is able to provide.

- The *Expedited Forwarding* service [146] or *Premium* service [212] is considered as a *virtual leased line* service. Therefore, the bandwidth cannot exceed a specific threshold, but the end-user can leave it idle or use it to the full extent of its capacity. The holder of this line should not perceive any influence of the presence or absence of other users. It can be used for real-time applications or mission critical traffic such as network control.

- The *Assured Forwarding* service [129] or *assured* service [50] provides an exception of a certain amount of bandwidth to the customers. This means, that bandwidth cannot be guaranteed, but packets are labelled with high priority. These packets have higher probability to be transmitted over the network and higher priority than best-effort packets. It is intended for non-real-rime interactive applications.

The DS service architecture relies upon *per-hop forwarding behaviours* [23], which represent the basic building blocks for QoS enabled services. This includes mechanisms such as the use of packet schedulers, classifiers, DS code point definitions (i.e., short marks), and traffic conditioners (i.e., meters, markers, droppers and shapers). Per Hop Behaviour (PHB) describes the externally observable behaviour of packets that is supported for a given traffic class. Therefore, the DS architecture defines several PHBs corresponding to the different service levels described above.

To summarise, DS is significantly different from IIS. First in the DS framework, there are only a limited number of service classes. Since resources are allocated in the granularity of class, the amount of state information is proportional to the number of classes rather than to the number of flows like in IIS. DS is therefore more scalable. Second, sophisticated classification, marking, policing and shaping operations are only needed at the edge of the DS networks. Core routers need only to implement behaviour aggregate classification. Therefore, it is easier to implement DS networks.

One of the key goals of the DS framework is to achieve inter-domain QoS interworking. This requires a way of communicating service level agreements (SLAs) between providers including the SLA definition itself and an appropriate protocol. While in traditional approaches SLAs are rather static and handled manually by operators and providers, in the last years the idea of an automated SLAs exchange according to the network state has been receiving more and more attention. For this purpose, *bandwidth brokers* have been proposed to implement such dynamic SLAs negotiation [212] (see Section 2.2.2).

The Multi-Protocol Label Switching   Traffic engineering is an iterative process of network planning and optimisation. The purpose of traffic engineering is to optimise resource efficiency and network performance [9], [362], [363]. For effective traffic engineering in IP networks, network operators must be able to control the paths of the packets. This requires some kind of connection in the connection-less IP networks. The Multi-Protocol Label Switching (MPLS) [264] scheme has been introduced by the IETF in order provide connections in IP networks with label switched paths (LSPs). MPLS is an advanced forwarding scheme: each MPLS packet has a header containing a specific label. An MPLS capable router, called Label Switching Router (LSR), examines the label and possibly the *experimental* field in the forwarding packet. IP packets entering and MPLS domain are classified and routed by an edge LSR based on a combination of information maintained in the router and carried in the packets header. More precisely, the MPLS label is used as the index to look up the forwarding table of the LSR: the packet is processed as specified by the forwarding table entry and the incoming label is replaced by the outgoing label and the packet is switched to the next LSR. Within this framework, the IETF also proposed constraint-based routing [59] and enhanced link state IGP [10]. Constraint based routing takes into account constraints such as bandwidth and administrative policy beside typical information about network topology. In this way, it is possible to compute longer but lightly loaded path better than the heavily loaded shortest path. In order for constraint-based routing to compute LSP paths subject to constraints, an enhanced link state IGP must be used to propagate link attributes in addition to normal link state information. Common link attributes include available bandwidth and link affinity, that is an administratively specified property of the link. In [364], a unified framework for providing QoS in the Internet that deploys MPLS, constraint-based routing, and enhanced IGP is proposed. These network layer mechanisms are integrated with the DS transport layer.

2.2 From Networking to Interworking Tasks

Network provider services comprise the end-to-end transport of bit streams (or data traffic) in specified formats over a set of routes. These services can be differentiated by quality guarantees such as speed, delay, errors, etc. *Networking* tasks include all the mechanisms (or component functions) that a network operator needs to implement in order to effectively provide these services. In detail, this includes three main categories:

- *Transport functions* group all the mechanisms associated with the transport of the information between remote locations using a shared transport network, e.g., multiplexing, error control, etc. They structure the client information into a data format, according to the precise needs of the underlying network, and they transfer the information across the network between the desired end points.

- *Control functions* are responsible for identifying data traffic and routing it to the appropriate end points within the transport network. They enable a user of the transport network to set up and take down connections within the context allowed by management functions, e.g., routing, signalling. The process of controlling how traffic flows through a network so as to optimise resource utilisation and network performance is called *traffic engineering*.

- *Management functions* include all activities associated with planning and design of the data transfer function, establishment of the context within which users can use the control functions (i.e., service management), maintenance of services, maintenance of the equipment in the network, control of network resources, etc.

While for the transport layer reasonably well established standards do exist (see Chapter 3 of [300]), the scope of control and management is very broad and many different approaches have been tried. Furthermore, the distinction between these last two layers is in several cases ambiguous, even if the ITU-T formally distinguished them in [250], [251]. In this context, a control and management functional architecture based on a set of implementation independent control management and functions is considered. Although our approach is illustrated by dealing with a specific service and network model (see Chapter 3), the aim is to define techniques broadly applicable (see Section 8.5). The emphasis is placed on the control and management functions required for the configuration of end-to-end services across different networks independently on the low level physical details specified by transport functions.

Within this context, *interworking* is conceived as a global process which includes both intra- and inter-domain control and management functions. A fundamental observation is that between distinct organisations, end-to-end connectivity becomes a function not just of physical connectivity, but also of intervening provider routing and management policies that are influenced by several important factors.

- *Heterogeneity.* What was once state monopolies controlling everything from end user access down to the copper wires has become several layers of competing firms (service providers, networks providers, retailers, brokers, etc.). Distinct networks can rely on different technologies and deploy different network management platforms. This implies heterogeneity of the information models used in different networks. In the best case, Telecommunication Management Network [140] (TMN) compliant domains, for instance, use standard TMN-X modules, which provide a rudimentary low level interface for synchronising the settings in routers and other elements at the boundary between distinct networks [113]. Even in this case, however, human operators are responsible for supervising and controlling the interaction.

- *Distribution of resources and control.* Network resources can be owned by many different authorities that need to be made to work together for supporting advanced services spanning several domains. This task is even more difficult for networks which aim to provide any kind of QoS guarantees, since individual providers are unwilling to release detailed information about the state, or even topology of their internal network.

- *Flexibility.* Currently, many aspects of the interaction between distinct networks are statically fixed by specific SLAs and many steps of the interaction are regulated by human operators via fax, e-mail, etc. This makes the overall interoperability process very slow (several weeks can pass before effective inter-domain network configuration changes take place) and inefficient.

Considering this complex and articulated scenario, a number of existing models for multi-provider connection establishment, ranging from traditional control plane solutions, based on signalling procedures, to connection management architectures, based on distributed computing technologies, have been studied. In the following, the focus is on those approaches relevant to the establishment of semi-permanent connections through the activity of the management plane that is the area most closely related the NPI approach.

2.2.1   Interworking in Connection-Oriented Networks

The most traditional approaches to interworking in connection-oriented networks, both circuit circuit switched (PSTN) and packet switched (e.g., ATM) systems, have been developed either within the TMN or the TINA frameworks. The central idea is the definition of standard interworking interfaces for exchanging limited information about network resource availability, topologies and routing policies. Appropriate signalling techniques built on top of these low level interfaces guarantee physical interconnectivity.

The Telecommunication Management Network Approach

The Telecommunication Management Network (TMN) has been defined by the ITU-T [140], [143] with the aim of providing a unified framework for facilitating the management of heterogeneous networks by organising, across a certain number of layers, the management of diverse communication components.

More precisely, the TMN architecture defines three main conceptual layers that address different needs to the organisation of a providers operations.

- The *TMN information architecture* refers to methods of representing network resources in managed systems as object-oriented abstractions which are observed, modified and controlled by managing systems via a manager/agent relationship.

- The *TMN functional architecture* refers to the breakdown of TMN network components into categories of functional blocks and classes of interconnections between them called *reference points*. This provides a way to describe the overall functionality of any TMN in a standard manner with a reference diagram.

- The *TMN physical architecture* covers the description of systems and means by which standardised communication interfaces are implemented in a TMN and the methods by which non-TMN interfaces are adapted to the TMN standards.

TMN standards define a logical layered architecture in which management tasks are broken into subsets and then partitioned into layers. TMN functions exchange management information by means of techniques defined in the X.700 series of reccomendations [249, 252]. These reccomendations incorporate two important concepts. First of all, data and the executable methods which access it are grouped into a fully encapsulated software object. The data can be manipulated only via the access methods provided in the object. Moreover, the *manager/agent* paradigm enables the hierarchical exchange of management information between systems, see Figure 2.4. A managing system assumes the role of *manager* for the purpose of issuing directives and receiving notifications, and a managed system assumes the role of *agent* for the purpose of carrying out directives and emitting notifications. A system that plays the role of manager to a lower level system may simultaneously play the role of agent to a higher level system, allowing for a cascading management hierarchy. The set of managed object classes and instances under the control of an agent is known as its management information base (MIB), an abstraction of network resources, properties, and states for the purpose of management. Each function in a layer of the TMN architecture represents itself and the resources it manages to the layer above by means of a managed object. This relies upon the collection of communication protocols called CMIS/CMIP. By making use of CMIS primitives, CMIP provides a set of capabilities such as information filter-

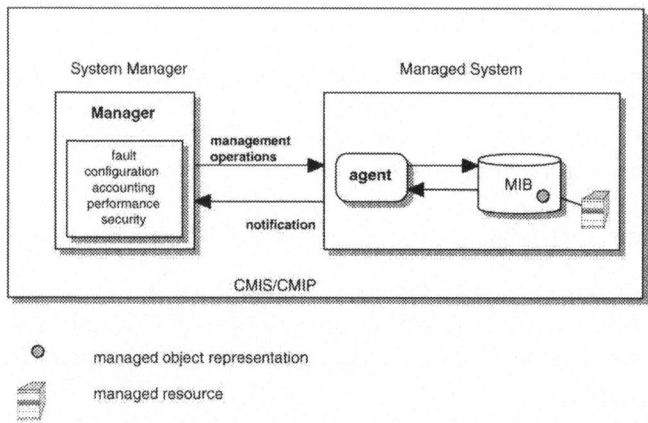

Figure 2.4: *The client/server paradigm which the TMN framework relies upon identified manager and agent roles by specifying which operations are allowed by system managers and managed systems.*

ing, asynchronous events report, network elements configurations, etc. The main possible services the use of the CMIS/CMIP set enables are listed in Table 2.1.

In the way the TMN architecture is defined there is a clear distinction between inter-TMN and intra-TMN communication. Most intra-domain communication takes place between TMN components that are in hierarchical relations from operations systems down to network elements. The communication along these lines is achieved over "Q" interfaces [253], [254]. When distinct operators need to establish connections crossing several administrative domains, they make use of specific predefined functionalities through the reference point, so called *X interface* [144]. An inter-operable TMN-X interface can exist between each physical instance of an operations system function (OSF) at a particular management layer within a single organisation and its peer instance within partner organisations and customers. For any management system that exchanges information across the TMN-X interface, the scope of management activities is defined by the specific data shared at networks' boundaries. The information model supported by each of the communicating TMN building blocks defines the range of effects of management operations inside every distinct network. The information model used to share a common view of available management information and functions at the TMN-X boundary is also called Shared Management Knowledge [145].

Several works have investigated various issues related to the specification and the implementation of the "X" interface. In particular, the PREPARE project [238] was set up as a part of the European RACE research and technology initiative to investigate the network and service management issues in the multiple operator and service provider context of a deregulated telecommunication market. The output of this work resulted in a number of contributions to the standard bodies

CMIS performative	Description of the service
M_CREATE	To create an instance of a managed object.
M_DELETE	To delete an instance of a managed object.
M_ACTION	To request an agent to perform an action.
M_SET	To modify management information.
M_GET	To retrieve management information.
M_CANCEL_GET	To cancel a previously requested invocation.
M_EVENT	To report an event about a managed object.

Table 2.1: *The basic CMIS/CMIP services that are enabled within the TMN framework between managers and agents.*

particularly in the area of requirements for specification of TMN-X interfaces and modelling broadband Virtual Private Network services [288].

Among other initiatives, the MISA ACTS AC080 Project [241] has developed an ATM/SDH network independent TMN-X interface for path-provisioning. One of the most significant contributions of this work is the definition of a unified framework for multi-domain network management that encompasses the heterogeneity of different underlying network technologies [101], [102].

The Telecommunications Information Networking Architecture Approach

TINA specifies an ubiquitous software platform for service logic, covering both service operation (or service set up) and service delivery (or service management) [323], [243], [117] that draws heavily on the TMN architecture. Figure 2.5 sketches out the main logical components in the TINA architecture. The transmission and the switching resources of the network are modelled as resource components. These components are used by service components, namely telecommunication, information and management services. *Telecommunication* services consist of the set of activities to transport bits of information between terminals. *Information* services handle information resources such as movies, sounds or documents. This includes the storage content, the visualisation, as well as auxiliary services such as billing and caching. *Management* services group fault, configuration, including connection management, accounting, performance and security management.

The overall business model of TINA consists of a set of actors in different business roles and reference points identifying the interactions between those actors, see Figure 2.6. The network operator (or more precisely the connectivity provider) sells its resources to the other stake-holders involved in the service architecture. The user, also called the consumer, interacts with third-party service providers in order to obtain the desired services. This interaction is usually achieved through a service retailer that helps the user find the service providers needed. The retailer is also in charge of establishing trust between both parties, notably for security and billing purposes.

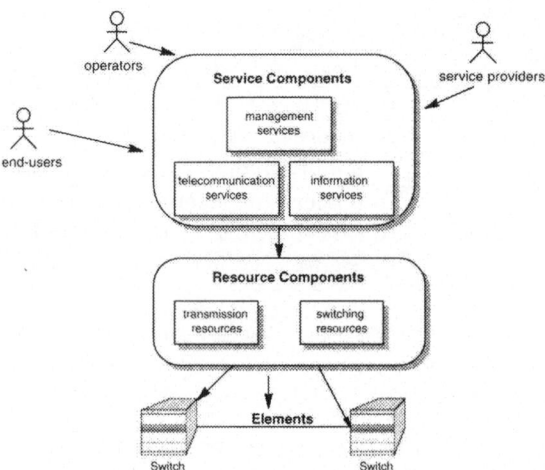

Figure 2.5: *The overall TINA architecture can be decomposed into two main logical distinct layers: the resource components and the service components blocks.*

In the TINA service architecture, the *Layer Network Coordinator*, LNC, is a resource component responsible for interconnecting termination points in a sub-network defined as a domain. If several distinct domains need to interconnect to each other, there is a direct interaction between distinct LNCs, [131]. Coordination is in this case achieved at the resource configuration management level. An alternative way to structure multi-domain connection management is the so called *federation* mechanism [227]. Instead of hierarchical relationships between computational objects (like when deploying LNCs), federation defines peer-to-peer relationships through the definition of standard interfaces. These kinds of interfaces are called in the TINA context Layer Network Federation (LNFed) inter-domain reference points. Domain federation is a static contractual relationship between providers and represents a pre-requisite for establishing services spanning different domains. Among various initiatives, relevant work has been done within the REFORM AC208 project [255]. Within this project, a TINA compliant architecture for multi-domain connection management functionality has been specified and implemented. The ACTS project VITAL [332] has focused on the interconnection of heterogeneous and distinct networks controlled by different providers through the use of federation. The first important outcome is the definition of a process for the configuration of resources needed to set-up and release connections. This enabled the evaluation of the feasibility of the TINA resource architecture, the enhancement of the TINA specifications and the detection of the principal aspects requiring additional work [203].

The main limitation of traditional approaches for multi-domain interactions is that in the face of today's growing and changing technology requirements low level interfaces, such as the TMN-X interface or the LNFed reference point, and

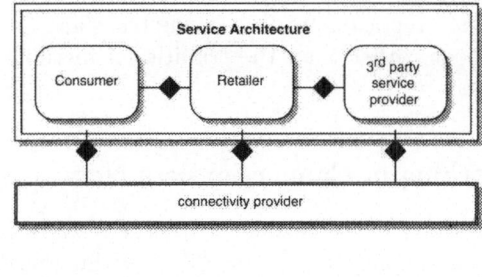

reference point

Figure 2.6: *The TINA business model consists of a set of participants in different business roles and reference points identifying the interactions between those participants.*

static domain federation are becoming increasingly inadequate. Neither the TMN-based framework nor the TINA approach provide functionalities for dynamic peer providers interactions. The basic primitives that distinct operators can perform on the common shared information are limited to dynamic updates reflecting the current state of pre-fixed resources according to pre-defined contracts. In the networking community, several organisations are currently working coming up with dynamic and flexible interworking solutions. Within the TMNForum [325], several industrial partners are working together in order to promote a higher integration and the cohesion of service and network management operations at both the intra- and the inter-domain levels. The SMART TMN initiative [214], for instance, enhances the industry standard TMN model and builds on the existing programs of the NMF by focusing on several critical areas such as:

- Automating and linking key business processes end-to-end;

- Integrating commercially available technologies;

- Enabling "off-the-shel" software development;

- Creating a set of common information and data elements.

SMART TMN seeks to extend the current view of TMN in four key ways.

1. Taking a holistic approach encompassing both internal interfaces (between a company ordering and billing systems, for example) as well as external interactions with other providers.

2. Embracing general purpose information technologies such as Java, CORBA and the Web, (and others where appropriate) in addition to the CMIP and GDMO technologies traditionally associated with TMN implementation;

3. Shielding the value of agreements from future technology shifts, by using a "protocol-neutral" approach to information modelling while also driving for consistent use (and reuse) of terms and models across processes;

4. Taking a more practical approach (using trials and development experiences) to development rather than the traditional method of bringing out paper specifications.

2.2.2   Interworking in Connection-less Networks

As mentioned in Section 2.1.2, the current Internet can be modelled as an arbitrary interconnection of autonomous systems (ASs). Within one AS, an interior gateway protocol is used, while between distinct ASs an exterior gateway protocol needs to be in place. Originally, an AS was essentially a routing concept, but it later has become a synonym of "administrative domain", which refers to the set of logic and network resources under the control of a single administrative authority.

Nowadays, the Internet consists of a large number of interconnected regional and national backbones provided by different Internet Service Providers (ISPs). Every ISP can control one or several ASs. In this context, routing protocols and data delivery across distinct domains have been extensively studied [261] [259], [74], (see also Chapter 2 of [76] for a good review), while a crucial aspect that has not been systematically addressed until quite recently is the feasibility of establishing and maintaining consistent dynamic agreements between distinct IPSs. Section 2.2.2 reviews relevant work that is in this area.

Exterior Gateway Protocol

Routing traffic between distinct domains in the Internet has been traditionally performed by applying the Exterior Gateway Protocols (EGP) [202]. EGP is defined for communication between the core Internet routers. These core Internet routers form the so called Internet routing backbone. Information is passed from individual source networks up to the core routers, which pass the information through the backbone until it may again be passed down to the destination network. In some cases, the connections between lower-level networks and the core routers are assigned and remain static; in other cases, EGP is used also for dynamically establishing these connections.

EGP does not use metrics and therefore cannot make true intelligent routing decisions. Routing updates sent to neighbouring routers at regular intervals contain network reachability information. Each router refers to those networks to which it is directly attached so that every router in the network can construct and maintain routing tables.

One of the main weaknesses of EGP is its incapacity in dealing with the routing loops that can occur in multi-path networks. Moreover, EGP routing updates are often very large and cumbersome. Finally, as mentioned earlier, intelligent routing decisions are not possible because the protocol does not support link metrics. For these major reasons, EGP has been gradually replaced in the Internet by the Border Gateway Protocol (BGP).

Border Gateway Protocol

BGP [185] is a protocol for exchanging routing information between gateway hosts
(each with its own router) in a network of autonomous systems. It is generally used
in the Internet as the inter-domain routing algorithm. The routing table contains
a list of known routers, the addresses they can reach, and a cost metric associated
with the path to each router so that the best available route (i.e., least cost route)
is chosen. Hosts using BGP communicate using TCP and send updated router
table information only when one host has detected a change. Only the affected
part of the routing table is sent. BGP-4 [260], the latest version, offers more
flexible routing policies by enabling administrators to configure cost metrics based
on policy statements.

BGP can perform three main types of routing:

- *Inter-autonomous* routing occurs between two or more BGP routers belonging to different autonomous systems. Peer routers in these systems use the
  BGP protocol to exchange information for maintaining a consistent view of
  the internetwork topology.

- *Intra-autonomous* system routing occurs between two or more BGP routers
  located within the same autonomous system in order to maintain a consistent
  view of the system topology. At this level, BGP is also used to determine
  which specific router will serve as the connection point for specific external
  autonomous systems. Therefore, the BGP protocol can provide both inter-
  and intra-autonomous system routing services.

- *Pass-through* autonomous system routing occurs between two or more BGP
  peer routers that exchange traffic across an autonomous system that does
  not run BGP. In a pass-through autonomous system environment, the data
  traffic did not originate within the autonomous system in question and is
  not destined for a node in the autonomous system. BGP must interact with
  whatever intra-autonomous system routing protocol is being used to successfully transport data through that autonomous system.

Bandwidth Brokers

The term *bandwidth broker* (BB) has been recently used in communication networks to refer to the entities that trade raw bandwidth between any two defined
points in the Internet. Their coordination may be centralised or distributed and
their functionality ranges from simple resource allocation tasks to complex trading
systems that include pricing and/or routing mechanisms. In more details, Nichols
et al. [212] proposed BBs for DS networks. The central part of their work is focusing on the global architecture mechanisms, packets marking, and integration of
other architectures like the IIS. In this framework, information exchanges between
distinct BBs can range from static agreements (i.e., classical peering) without any
signalling to dynamic setup even for a single changing flow.

Clark and Fang [51] describe the *allocated-capacity* framework which uses a single bit to differentiate services. Service allocation profiles are based on traffic specifications, geographic scope, and probability of assurance. At every network access point actual traffic is metered and tagged according to the conformance or not to the pre-fixed contracts regulating multi-domain interactions. This model charges only traffic that causes congestion (other traffic is covered by flat-rate charges). With this approach, it is not possible for the distinct domains to dynamically renegotiate the traffic profile.

Dynamic agreements for multi-provider interactions within DS networks have been studied and proposed by Fankhauser et al. [79], [80]. Dynamic peer-to-peer relationships are established by software entities called *SLA traders*. These entities follow market-based principles to decide which contract is more beneficial to the organisation they are acting on behalf of. In particular, SLA traders compare the offers made by neighbour providers and select the most profitable ones. The selection of specific peer services creates competition among providers and integrates path selection based on service level and destination. SLA traders act locally but they exchange information to build a global view of the DS networks through distributed routing algorithms. Experimental analysis validate this approach for highly connected areas of the Internet where a significant amount of traffic is forwarded [78]. In the SLA-trader based environment, every distinct provider domain is modelled as a single node, therefore, intra-domain resource allocation issues are not taken into account. The case of multiple connections between distinct ASs is also not covered.

The Internet2 initiative [137] features an overlay network, the QBone testbed that supports different levels of service quality by initially implementing the DS approach. Here, a bandwidth broker responsible for allocating four main different service levels is currently under development and the implementation will be fielded in the CA*Net II [321]. CA*Net II is conceived as a separate, high-speed Internet system which will help further development of Internet applications and related communications technology. This initiative is intended to connect universities, research organisations and certain businesses. The fundamental idea is, in a first phase, to implement basic features of the DS approach and later to allow dynamic SLAs trading based on the service destination (network prefix). There are currently several implementations of BBs [2]. Telia, for instance, developed a software broker that is intra-domain centred, cooperates with OSPF [3] to learn the topology of its domain and uses SNMP [4] [40] for network setup and router interactions. The emphasis in the approach of Telia is on the protocol processing performance between these software brokers [287].

---

[2]More details about the Bandwidth Brokers Projects can be found on-line at the following URL:
http://www.internet2.edu/qos/qbone/QBBAC.shtml

[3]OSPF [205] is one of the most common routing protocols used in the Internet today.

[4]SNMP is is the protocol governing network management and the monitoring of network devices and their functions in the Internet world.

Lazar, Semret et al. [175], [291] focus on dynamic pricing mechanisms in DS networks and consider the stability and the consistency of bandwidth allocation across several domains. For this purpose they make use of bandwidth brokers acting as mediators between consumers (or users) and network operators (or bandwidth sellers). The *progressive second price* auction mechanism is proposed for coordinating users, service bandwidth brokers and raw bandwidth sellers. Their game theoretical approach is proven to be incentive compatible and efficient. In [292], Semret shows that stability can be achieved even for different services classes and service levels affecting each other, *thus the good news is that dynamic market-driven partitioning of network capacity among services appears sustainable.* However, he observes instabilities at the peering level, or macro level (i.e., between distinct providers), even in small networks (three network nodes are used in his simulations) and tight provisioning. In Section A.1, we discuss in more details the relevant aspects of this approach in relation with the pricing process.

2.3 Interworking Evolution and Innovative Approaches

Traditional approaches for solving the MuSS problem heavily rely upon the human control. At the intra-domain level, operators control low level network components through the use of tools and/or application programming interfaces. At the inter-domain level, human controllers directly supervise low level interfaces (e.g., TMN-X interface) and usually interact between each other via phone, fax, e-mail, etc. This makes the overall interworking task very slow and quite inefficient (for instance, several weeks can pass before effective inter-domain network configuration changes take place). Inefficiency is mainly due to the difficulty for human operators to consider all aspects which complicate the interworking process:

- *Market Liberalisation.* The increasing number of actors (content providers, added-value service providers, brokers, etc.) with different roles than those of the traditional Telecom operators. With market liberalisation the difficulties of pan-industry coordination clearly increase in line with the number of actors involved. Competition is fierce and has kick started an industry wide drive for efficiency.

- *Rapidly changing technology.* The number and diversity of deployed network technologies is continually growing. This diversification is creating a complex heterogeneous network infrastructure and serious technological challenges in providing uniform and coherent services. There is often not enough time for the industry to develop guidelines before those guidelines are already obsolete. Standards bodies (such as the ITU, ISO, ANSI, ATM Forum and IETF) are having to catch-up with common practice rather than setting the agenda.

- *Increasing flexibility in usage requirements.* With market differentiation and increasing customer demand comes a need for flexible service use. Therefore, the need for an efficient mapping between intra- and inter-domain management and control aspects is more urgent. Networks need to be adapted to provide what customers are requiring, cope with fluctuations in usage and handle the introduction of new multimedia services (such as video, audio, Internet telephony and e-commerce related communications).

These factors together are combining to produce very complex network architectures and requirements that can be considered as the main business drivers which are generating potential need for agent based solution [351], [124]. In parallel, important technology push factors have been reinforcing an increasing interest in agent technology for several kinds of networking applications. Until recently, many agent applications have remained nothing more than small pilot projects in the research laboratory. One key reason behind this is that the necessary network architecture for agent application was just not available, but this has been changing.

- The utility of mobile agents and mobile code for network control has been a recurring theme since the early 1990s. This paradigm is now beginning to gain wider acceptance in the Communication Networks community [12], [32]. In addition, several projects have produced relevant results for flexible service provisioning as well as automatic feature updates in terminals and network nodes [86]. Consequently, the likelihood that agent capable platforms might be supported by future networks is increasing.

- Researchers in the field of Active Networks [322] argue for programmable network which can receive and execute code on time scales down to single packet arrival. Programs can be down-loaded to a router using a "back-door" mechanism or injected into the network in the headers of individual data packets. Either way, this type of programmable networks would greatly increase the scope for the application of agent based network control services into the network infrastructure[5].

- Ongoing standardisation efforts within bodies such as OMG and FIPA are providing standard interaction mechanisms for agent based software. These efforts to provide interoperability for agent applications are a key factor in enabling the use of agent technology for a large range of tasks - including in network related applications.

- There are now a large number of toolkits available which enable developers to quickly build multi-agent systems. In addition, increasingly, toolkits are also including reasoning capabilities and GUI management tools for agents developed.

In addition, the FIPA community envisaged the network management scenario as one of the most interesting and important agent application fields [87].

---

[5]Some CISCO Systems routers in fact already include Java Virtual Machines.

However, FIPA only gives the basic guidelines about the functionalities that software entities should provide. The details of how to integrate different agents and improve the service provisioning process with intelligent techniques, i.e., coordination, collaboration, etc., are left to application developers. This triggered several initiatives that tried to address many of these issues. In particular, based on the FIPA Reccomendations [87] the European Project FACTS[6] developed a fully agent based architecture for the provision of timely and highly personalised suggestions for TV viewing, using a natural language interface and synthetic character Agent. FACTS used a FIPA compliant architecture to automate all levels of the information supply chain demonstrating the feasibility of agent-based solutions in existing network infrastructures.

Above all, several initiatives such as the AgentCities [1] initiative and various Information Society Technologies (IST) European Projects [320] are currently grouping numerous organisations from both the academic and the industrial worlds to investigate the commercial and research potential of agent-based applications.

2.3.1 Agents in Networks: a Short Review

The idea of distributing communication network control and management tasks by making use of 'smart', 'cooperative' and 'autonomous' entities in network infrastructures has received considerable attention from both the DAI and the networking communities [194], [311], [58], [235], [351]. For instance, Weihmayer and Brandau proposed cooperative distributed solutions to communication network management [343]. In particular, they propose the TEAM-CSP system for customer network management. TEAM-CPS agents control a private network and interact with public network management. The emphasis in this framework is on the problem solving activities that agents have to perform. In order to make sense of the specific domain, they maintain in a database beliefs and commitments about the real world state. So and Durfee developed a distributed network management system called Distributed Big Brother (DBB) that makes use of different DAI techniques such as contract formation, organisational structuring, election for role assignment, and hierarchical control [309]. The resulting cooperative problem-solving infrastructure is deployed for network monitoring and management. LODES is another multi-agent system developed for diagnosis and monitoring of TCP/IP based LANs [315]. This system consists of a set of expert systems that control different aspects of a network and cooperate with each other in order to find faults that occur in it. Different network segments are monitored by different agents for detection of faults. While in DBB a hierarchical approach is followed, in LODES agents are organised into peer-to-peer relationships. In [57], a multistage negotiation protocol, which extends the functionality of the contract net protocol [308], [62], is presented (see Section 2.4.1 for more details about the contract net protocol).

---

[6]More details about the European Union ACTS Project FACTS can be found at http://www.infowin.org/ACTS/RUS/PROJECTS/ac317.htm

The protocol is applied to the task of restoring paths in communication networks. Distributed agents monitor different parts of the network, construct partial plans to accomplish common goals and deploy the multistage negotiation protocol to integrate these partial plans into a global one. Many other examples could be described, however, since an exhaustive review of the use of agents for communication networks and management is out of the scope of this volume, for more detailed accounts of previous work, interested readers are referred to good surveys such as [167], [344], [123], several papers in [124] and previous proceedings of the IATA workshops[7]. In the following, the attention is on relevant agent-based paradigms for routing, managing and controlling multimedia traffic over multi-provider networks.

2.3.2   Related Agent-Based Approaches for Multi-Provider Networks

As mentioned in Section 2.3, the European Project FACTS investigated the application of autonomous software agents for service provisioning in multi-provider scenarios. Within this context, negotiating agents act on behalf of network providers (i.e., owners of the network resources) for defining offers that are proposed to service providers [84] (i.e., providers of networking services that do not necessarily own the underlying physical network resources). A service provider acts as a broker negotiating with a number of network provider agents and selecting the provider offer (or the combination of offers) that is the most suited to provision the resources needed for a certain end user demand. However, negotiations between peer network providers have not been explicitly considered in the FACTS Project [83], even if the formal negotiation model developed by Faratin [82] could be eventually applied to this kind of framework by directly mapping the set of negotiation tactics to network resource management and control mechanisms. This is where this work brings more explicit contributions.

In [122], Hayzelden and Bigham propose a multi-agent based approach to the problem of making a logical network resource configuration and adapting it to customer utilisation. The agents have goals derived from different quality metrics, which the network has to provide. In ATM networks a well designed Virtual Path Connection (VPC) overlay network tries to maximise the probability of being able to accommodate connection demand within the control plane for that particular time frame. They assume that there already exists a pre-defined VPC route topology in operation. In the NPI approach, intra-domain routes are not predefined, but computed on-demand. The distributed agent control architecture (Tele-MACS) illustrated in [122] couples planning agents with reactive agents that can adapt the network resource configuration to comply with changes in user demand. A main difference between the NPI framework and Tele-MACS is that in

---

[7]For the *Intelligent Agents for Telecommunications Applications* (IATA) Proceedings see [135], [271], [272].

the Tele-MACS context the network scenario is considered a closed system, where the network operator has full control over the network elements, such as the routing tables and switching capabilities. In our work, since we consider end-to-end demands spanning networks controlled by distinct operators, network elements cannot be managed and configured only by a single provider. More recently, in the context of the European Project IMPACT[8], Bigham et al. describe a more complex multi-agent system managing admission control in ATM networks [18], including routing, dynamic bandwidth management, etc. The focus is on interactions between end users and service providers and between service and network providers. These interactions are modelled as first-price sealed-bid auctions. Since end-to-end paths, namely ATM virtual paths, are assumed to be fixed (only the amount of available bandwidth on them is varying), no dynamic negotiation is considered between distinct service providers and/or network providers. A Connection Agent acting on behalf of the user conducts an auction where individual Resource Agents inside a specific service provider domain submit bids. IMPACT does not aim to address the management of a service provider or a network provider domain and their dynamic interactions.

Gibney and Jennings describe the design and the implementation of an agent market-based system for call routing in telecommunication networks [108]. A two-level economy model has been proposed. At a first level, 'call agents' acting on behalf of end users buy paths by interacting with 'path agents' through a sealed-bid auction. At a second level, 'path agents' buy capacity units by 'link agents' through first-price sealed-bid or Vickrey auctions. The most interesting idea is that by using a two-level economy it is possible to realise call admission control in the same framework as the network management function. However, although experimental results validate this approach vis-a-vis of static routing there are some very restrictive assumptions that we believe cannot hold in realistic networks. First of all, mainly for scalability reasons, it does not seem feasible to allocate agents for every link and for every possible path resulting from all the possible combinations of those links (even if the authors limit the choice to three known and static paths). Furthermore, path agents are supposed to buy the links that are necessary for forming end-to-end paths by sequentially interacting with different link agents within distinct auctions. This is a typical example of combinatorial problem for which sequential and/or parallel different auctions have been proved to have serious limitations [224], [278]. In our framework, end-to-end paths are created through bilateral negotiations that take place between peer providers involved in the allocation. These interactions run in parallel. The pre-filtering of local solutions (routes) that are not consistent with boundary constraints (done via CSP based techniques) eliminates the need of sequential bargaining (see Section 5.1 for more details). The inter-dependency of different and parallel negotiation processes can be automatically taken into account by propagating constraints.

---

[8]More information about the ACTS Project IMPACT can be found at http://www.infowin.org/ACTS/rus/PROJECTS/ac324.htm

Stamoulis et al. [314] propose software agents negotiating for telecommunication services (e.g., video on demand) on behalf of the user. Their approach is
very user-centric: the objective is indeed the definition of an automated framework
allowing the selection of the best of the service combinations offered by multiple
service retailers. There is no assumption about the tariff structure that service
providers adopt. The fundamental difference with our work concerns the way different providers are assumed to interact. In the paradigm discussed in [314],
long-term and static bilateral agreements exist between neighbour networks to
establish the amount of aggregated traffic that is supposed to be routed through
pre-designed access points. A 'bandwidth broker' agent controls the optimisation
of intra-domain resource allocation, so that the inter-domain aggregated level of
traffic is respected, but there is no way to influence or dynamically modify the
inter-domain traffic routing.

In [208], Nakamura et al. introduce agents for pricing and accounting mechanisms for connection-less services requiring QoS guarantees. In their framework,
a network access agent (NAgent) receives a communication requirements from an
end user. The NAgent computes the end-to-end route until the destination node,
i.e., all intermediate nodes and access points to be crossed are pre-selected, claims
the communication fee, and divides it among the providers from a position independent of the providers. This implies that every provider submits a detailed view
of its topology, offered QoS and pricing system. This kind of approach suffers from
typical problems of scalability and robustness of all centralised solutions. Secondly,
no dynamic negotiation about different local routes can take place, therefore it is
not clear if utility and resource utilisation can be dynamically optimised inside
every network.

2.4   Multi-Agent Systems and Coordination

Software agents represent the key technology that has been adopted in this volume
in order to model and define the various participants to the MuSS process and
their relationships. However, agents per se are not enough. The integration of
powerful coordination mechanisms is essential to define an effective system able to
regulate interactions of autonomous self-interested entities controlling distributed
and private resources. In the following part of this Chapter, we review the basic
building blocks upon which coordination mechanisms have been defined in the NPI
framework, namely *negotiation* and *distributed constraint satisfaction* techniques.

2.4.1   Negotiation Techniques for Coordination

A multi-agent system (MAS) is a system containing multiple agents that work together in order to solve problems that are beyond their individual capabilities [219].
In the MAS field, research is primarily concerned with coordinating a collection of
autonomous and individually motivated computational agents [345], [346], [163],

[266], [284]. Among several existing approaches including organisational structuring, contracting, planning and social laws (see [340] and [217] for more exhaustive surveys), decentralised models of *negotiated* coordination are primarily considered.

The main motivations for using negotiation-based methods for coordination come from in depth investigations undertaken by economists about the problem of coordinating multiple agents in a societal structure [159], [211]. The key point is that economics offers mechanisms that produce globally desirable results, avoiding central coordination, and imposing minimal communication requirements [350]. Furthermore, negotiation represents one of the most common and natural mechanism for conflict resolution in many real applications. For instance, several works have recently proposed negotiating agents in various e-commerce applications (see [277], [276], and several papers in [204] and [216]). The Contract Net Protocol (CPN) [308], [62] is one of the most known coordination techniques for task and resource allocation making use of a market-like paradigm. Agents can take two roles: managers and contractors. A manager sends out a request, i.e., a contract announcement, to a number of potential contractors in the market to provide a specific service. The contracting agents submit bids in response to the announcement and the manager selects the best bid, i.e., the contract is assigned to the contractor who submitted the highest bid. Although many researchers consider this approach as a negotiation based paradigm, many others classify the CPN as a standardised coordination method since there is no possibility of bargaining between the agents and there is no two-way agreement [219], [112]. In the MuSS scenario, the CPN approach is unable to provide a coordination mechanism capable of dealing and resolving the possible conflicts arising between negotiating agents.

Among relevant work done in the negotiated coordination area, a first possible distinction is between the two following sets of models:

1. Coordination models for *societies* of agents such as voting, general equilibrium market-based mechanisms and auctions.

2. Bargaining models for *individual* negotiating agents including game theoretical models, heuristic approaches and normative models of bounded rationality.

Since the work developed within this context concerns bilateral negotiations, a brief overview of the most common coordination models for agent societies is given, concentrating on the one-to-one bargaining models. Interested readers are referred to more exhaustive surveys such as [174], [282], [184], [150], Chapter 2 of [366], and Chapter 3 of [82].

2.4.2   Societal Approaches

In the area of negotiated coordination, there exists a set of models that provide mechanisms for coordinating societies of agents.

- *Social Choice* or *voting* models [4], [293], [294] establish a specific mechanism to which all agents submit their input. The output that this mechanism determines as a function of the submitted inputs is a solution for all the agents and bind all of them. In voting, the protocol designer is assumed to be willing to enhance the social good. The most known and common example of social choice rule is the majority voting. The main objective in social theory has been to define a choice rule that orders possible social outcomes based on individual preferences of those outcomes. Arrow's impossibility theorem [4] raises serious issues about the rationality of the overall solution that can be obtained with social mechanisms. Although each agent may assign utilities or preferences to different alternatives which are perfectly rational, the outcome of the selection process may not reflect majority opinions or transitive orderings. However, several efficient social choice solutions have been proposed [114], [52], since in many real settings the individual preferences are restricted in a way that the conditions of the Arrow's impossibility theorem are not valid any more. Good reviews of the fundamental principles of social choice theory can be found in Chapter 5 of [166] and in [282].

- *General equilibrium* theory defines a micro-economic approach for computational markets, i.e., markets in which software entities, producers and consumers, trade a finite number of commodities between each other. The amount of each commodity is assumed unrestricted and each commodity is assumed arbitrarily divisible. Furthermore, agents are considered as price takers, i.e., the market is so large that no single agent action can influence the prices. This theory has been adopted in several application domains such as flow routing in a network [349], configuration design [350], power load management [367], bandwidth allocation on ATM networks [170] and adaptive QoS control for multimedia applications [365] (other examples can be found in [54]). The market reaches a general equilibrium when a set of prices (one for each commodity) is found such that supply meets demand for each commodity and agents maximise their use of the resources at the current price levels [198]. Such an equilibrium is always Pareto efficient. Equilibrium, i.e., an efficient joint solution accommodating all market participants, can be found without centralising all the information by deploying two very known algorithms: the price tâtonnement and the quantity tâtonemment processes (see Chapter 2 of [366] for a good review of these algorithms and their characteristics). The *tâtonnement* process is an iterative mechanism: at each iteration, an auctioneer sets a vector of prices and all the agents have to declare how much they are willing to buy and/or sell of each commodity at the current declared price. Based on this information, the auctioneer comes up with the new vector of prices. Two important issues in general equilibrium theory that have been extensively discussed are the existence and the uniqueness of the equilibrium [198], [330].

More attention is devoted to *auction-based* approaches since several relevant works in the area of communication network management have deployed this kind of techniques to achieve coordination of distributed agents.

Auctions

The concept of auction is very common and almost everybody is familiar with such kind of trading mechanism. Nothing very exotic seems to happen when an auctioneer offers Leonardo Da Vinci's diary, some people bid, and when everybody is silent, the diary is sold to the person who submitted the highest bid. However, the simplicity of such a mechanism is only apparent. Many different aspects must be considered in order to perceive the complexity which is behind an auction.

First of all, there are qualitatively different auction *settings* depending on how bidders valuate a good. In *private auctions* the valuation is not dependent on external factors and it is based on private bidder information. This is typical when the buyer is acquiring goods for personal use. If the valuation of bids is decided not only upon private considerations, but also upon valuations of others bids we have *non-private* auctions. If the value is partly dependent on other values the auction is also called *correlated value* auction. Whether the value is completely dependent on other bid valuations the process is identified as a *common value* auction.

Second, several kinds of possible auction protocols exist. An *auction protocol* establishes the order in which prices are quoted and the manner in which bids are tendered. There are different ways of classifying auctions, but generally experts agree that there are four major one-sided auction protocols (see also [275] and [168] for good reviews).

1. *Open-cry* or *English* or *ascending-price* is probably the most known auctions mechanisms. It is commonly used to sell art, wine and numerous other goods. We borrow the Paul Milgrom [226] definition:

   > Here the auctioneer begins with the lowest acceptable price (the reserve price) and proceeds to solicit successively higher bids from the customers until no one will increase the bid. The item is 'knocked down' (sold) to the highest bidder.

   All the buyers can hear bids submitted by rivals, which is easy in physical auctions, but in electronic environments this means that several minutes or more can be allowed for the response. Usually, the auctioneer displays the last highest bid on a Web page [9] or in agent-based systems the auctioneer can send back to every auction participants a message to inform about the current auction state [215], [34].

---

[9]See, for instance, www.ebay.com, auctions.yahoo.com, www.charityfundraiser.com and many other on-line auction houses.

2. *Dutch* or *descending-price* auction, which is the technique used (beginning in the Netherlands and then in many other countries) to auction perishable goods such as fish, flowers, etc. The auctioneer starts with a very high asking price that is progressively lowered until a buyer claims an item (or how many items he will purchase at the current asking price). If multiple units are offered, the first winner takes his prize and pays his price and later winners pay less. Examples of multi-agent systems supporting Dutch auctions are the Fish-Market simulator [263] and the MACH simulator [34]. Although such an auction is not very common in real (i.e., non-electronic) market-places, simulations can be found on the Internet [10].

3. *First-price sealed-bid* auctions have two main distinct phases: a bidding period in which all potential buyers send their bids (only one bid per bidder is allowed) to the auctioneer, and a resolution phase in which the bids are evaluated and winner/s are declared (sometimes the winner identity is not announced to all the participants). Each bidder does not know anything about others bids. It is possible to distinguish between two paradigms: the *first price*, and the *discriminatory* auction. In the first case a single item is offered and the highest bidder wins and pays the amount he bid. In a discriminatory auction there is more than one unit for sale, sealed bids are sorted from high to low, and items awarded at highest bid price until the supply is exhausted. In this latter case, winners can pay different prices. A single round first-price sealed-bid auction lacks of competition among the participants, but multiple round auctions can stimulate more antagonism. In a *multi-round* sealed-bid auction there is a deadline for each round of bids: at that deadline either the auction is closed or the bids from all participants are publicised and a new round of bids is solicited.

4. *Vickrey* or *second-price sealed-bid* auction[11]. Like in first-price sealed-bid auctions, each bidder submits a bid without knowing anything about others bids. The highest bidder wins, but at the price of the second highest bid.

   Vickrey auctions have been used in auctions among humans [268], [267] in the former Czechoslovakia to refinance credit and in Guinea, Nigeria, and Uganda for foreign exchange. In electronic auctions they have been widely studied and deployed (a good list of references is given in [274], [280]).

Another important auction paradigm, that has been used in U.S. financial institutions for over a hundred years, is the *double auction* protocol. In this auction both buyers and sellers send their bids to the auctioneer. Bids are then ranked highest to the lowest to generate demands and supply profiles. From those profiles, selling offers (starting with the lowest) are matched with demand bids (starting with the highest).

---

[10]http://www.mcsr.olemiss.edu/~ccjimmy/auction

[11]The uniform second-price auction is commonly called Vickrey auction. William Vickrey, winner of the 1960s Nobel Prize in Economic Sciences, first classified this kind of auction.

The third important aspect to consider after having defined the auction settings and the protocol is the specific *auctioning strategy* that participants decide to adopt. From the seller perspective, the main decision concerns the auction protocol, which means the seller has to predict the behaviour of bidders. In addition, a seller can influence auction results by revealing information about the good: an optimal strategy is to reveal information, since in general the more information a bidder has, the more a price moderation effect of winner curse is reduced [275]. A buyer makes an estimation of his own value of the good and evaluates what other bidders are likely offer. The *revenue-equivalence theorem* [198] demonstrates that, under the assumption of private value, all four basic auction types produce the same expected value to the auctioneer when bidders are risk-neutral and symmetric. However, this theoretical results do not apply in many real situations. From the bidder perspective the kind of auction protocol has the effect of revealing more or less information in the course of the auction. Bidders must decide the maximum amount they will bid, based on their own valuation of the good and their prior beliefs about the valuations of other bidders. If useful information is revealed during the auction, bidding strategies can take into account this additional input.

Smart auction principles and bidding strategies have been analysed and studied in various ways, in order to maximise the revenue coming from this kind of trading process. The bibliography goes from more theoretical and general aspects concerning all kind of auction [41], to very specific studies related to the use of an auction paradigm for selling specific products, such as wine [6], second hand automobiles [176], construction projects [334], etc. Nowadays, electronic auctions are becoming more and more popular due to basically two important factors. First of all auction mechanisms are better known. Deep studies of how to optimise auction protocols and strategies promise efficient results also in very complex settings (see for instance [224], [278], [132]), providing a distributed and autonomous way of solving task and resource allocation problems in multi-agent systems (see Section 2.4.2). Second, the success of the Internet and the development of e-commerce give to auctions a sort of 'natural environment', where a huge number of potential customers and suppliers can settle business transactions in a shorter time [168].

Auction-based Frameworks for Communication Networks   Many researchers are focusing on several auction-related topics and/or use auction mechanisms to achieve coordination of agents societies in the networking field. The two most significant areas for this work for which an extensive use of auction mechanisms have been proposed are *resource allocation*, including resource configuration, and *pricing* network components and/or services.

- *Auction-based Resource Allocation* Miller et al. [201] developed a system that allocates bandwidth in real time for a video delivery system. Their *discovering* system uses auction-based mechanisms inside an ATM network, in order to allocate network bandwidth and CPU time to application programs. *Auctioneers*, responsible for allocating ATM virtual circuits, and *deliverator* agents, managing virtual circuits, update the allocation and the pricing of bandwidth in real time responding to changes in demands and user preferences. The main limit of this system is the way the auctioneer is supposed to work. This centralised entity requires a global model of the network.

Gibney and Jennings [108] describe a system architecture which allow agents representing different network resources to coordinate their resource allocation decision. An auction-based approach is used in order to route calls through a network over time. There are *links agents*, which sell link bandwidth (link market), and *path agents* that act as both buyers of link resources and sellers of path resources (path market). Finally, some *caller agents* bid for path resources. The type of auction that is used for both markets is a *double* auction: buyers and sellers place bids and offers simultaneously on the market. The market matches the bids to the offers, so that prices and allocations of goods are established. In the variant of the double auction that is used in this framework, for each trading round buyers and sellers send bids and offers for known allocations of goods. This scheme is particularly rapid since the resources and payments are coupled with the offers and bids respectively and there is no negotiation overhead. However, some important scalability problems arise when considering to integrate this approach in real networks (see Section 2.3.1 for more details).

- *Auction-based Pricing* Auction mechanisms have also been widely used in the communication network community for dynamically establishing prices. MacKie-Mason and Varian [189], [190] use a Vickrey auction to allocate bandwidth in computer networks. In particular, they investigate how to use market-based techniques to price congestion in real time in the Internet. In their *smart market*, users include a bid in each packet. At congested routers, packets are prioritised based on these bids. In case of congestion, packets containing the lowest bid are discarded first, and accepted packets are priced at a rate determined by the highest bid among the rejected packets. The cost of carrying each packet is thus related to the marginal value (represented by the bid) of the traffic which has been pushed out. At the equilibrium price the user willingness to pay for additional data packets equals the marginal increase in delay cost generated by those packets. In order to make the scheme incentive compatible, users are not charged the prices they bid, but rather the bid of the lowest priority packet admitted to the network. Since the granularity of price setting is at the packet level, the smart market model is expected to have a high transaction overhead. Moreover, several technical challenges have to be met in order to implement this scheme in TCP/IP networks (see [306] for more details). Bandwidth auctioning, rather than per-packet auctioning, has also been proposed in [188], [81] and [256]. For instance, Lazar and Semret propose in [175], [291], [292] the *progressive second price* (PSP) auction mechanism for allocating variable shares of bandwidth among multiple users. The PSP mechanism achieves incentive compatibility, stability and efficiency, i.e., at the equilibrium allocation maximises total user revenue (see Chapter 3 of [290]), under the assumption of elastic demand analysed as a complete information game. This assumption is however very strong and not always valid in real settings where users cannot be considered as simple price takers.

Despite the apparent simplicity of auction based mechanisms, the integration of these kinds of coordination mechanisms in real networks on top of the existing control systems and signalling techniques does not seem to be a feasible solution at least in the short-medium term (see also Section 5.1.1). The main drawback arises from the need of reserving resources for the overall duration of an auction eventually for several incoming demands. This clearly becomes a combinatorial problem. Auctions can rather play an important role when *surplus* resources are available to be offered to the market. A concrete example is given by the band-X auction site (www.band-x.com) where network capacity, leased and owned clear channel circuits, dark fiber, wavelengths and ducts are offered.

2.4.3 Bilateral Negotiations

Coordination by means of negotiation between individual bargaining entities has been the central focus of a number of different areas such as economics [20], [19], in particular, game theoretic models have been considered as the basis of many computational systems [21], [376], [266], [335], social psychology [244], social welfare theory [5], [265], [105], marketing [61], operational research [301], organisational theory [38] and many DAI works.

In this area, the models in which coordination mechanism of market institutions is replaced by individual bargaining are subdivided in three groups:

- *Game theoretical models*: One of the earliest applications of game theory was to the study of market behaviour. Osborne and Rubenstein [221] present the theory of bargaining and its application to behaviour in markets. The idea is to focus on *monopsony* (one buyer and many sellers) and *bilateral monopoly* (one buyer and one seller) scenarios (see Chapter 15 of [166]). There is a major distinction between *axiomatic theories* or cooperative models and *strategic bargaining theories* or non-cooperative models. The former class includes approaches in which agents make binding agreements between each other before actually negotiating (pre-play negotiation). In addition, desirable properties for a solution, called axioms of the bargaining solution, are postulated and then outcomes that satisfy these properties are searched [107]. In non-cooperative models agents try to maximise their own profit without considering others benefit and the outcome is based on an analysis of which of the players strategies are in equilibrium [239]. The main limitations associated with the use of game theory when applied to automated negotiation in real applications, as Nwana critiques [217], are that:

  - Agents are assumed to be fully rational and acting as utility maximisers
    using predefined strategies;

  - All agents have knowledge of the payoff matrix and therefore full knowl-
    edge of other agents preferences.

This means that no computation is required to find mutually acceptable solutions within a feasible range of outcomes (infinite amount of computational
resources). Interesting issues concerning the application of game theory to
automated negotiation are discussed by Binmore and Vulkan in [22]. They
underline the kind of changes that are required to existing theoretical models
in order to overcome current difficulties and opportunities.

- *Heuristic approaches*: A cost is associated with both computation and decision making and more realistic assumptions about rationality of the players
  and availability of computational resources are made [150]. In particular, every agent decision making process is modelled heuristically during the course
  of negotiation. The selected protocol does not prescribe an optimal course
  of action, thus, good solutions rather than optimal ones are determined.
  In some cases, heuristic approaches are derived by game theoretical models by relaxing some of the fundamental assumptions about rationality of
  agents or amount of available computational resources. The central focus is
  in modelling heuristically the agent decision making process. Relevant work
  in this area is presented in [165], [319], [286] , [13], [85], [375], [133]. Heuristic
  methods have specific drawbacks vis-a-vis of game theoretical approaches:

  - Because of limited rationality and a limited amount of computational
    resources, sub-optimal outcomes can be selected.

  - Since it is normally impossible to determine a priori how negotiating
    agents will act during the negotiation, these models need extensive em-
    pirical and/or experimental validation.

Argumentation-based approaches represent a particular class of heuristic
methods in which additional information is exchanged "over and above proposals" [150]. Through argumentation negotiating agents exchange proposals
and counter-proposals accompanied with arguments which summarise the
reasons why the proposal should be accepted or not [304]. The argumentation is persuasive since the exchanges have the potential to alter the mental
state of agents involved in the negotiation process. This means that the negotiation space can dynamically change. Several works fall in this area. For
instance, in the case of labor negotiation, Sycara [318] presented a model
of negotiation that combines case based reasoning and optimisation of the
multi attribute utilities. This model has been further refined in [164] and
extended by Parsons and Jennings in [225].

- *Normative models of bounded rationality.* Like in heuristic models, a cost is associated with computation and decision making. However, for this class of approaches the aim is to establish the best strategy that agent can use given an interaction protocol. As expressed by Sandholm in [281] "a formal character of rationality limitations allows for the formulation of normative (prescriptive) theories of interactions of self motivated computationally limited agents". In [281], each agent has to pay for the computational resources (CPU cycles) that are used for deliberation. Then, a domain cost is associated to the total amount of computational resources to the problem (in the vehicle routing problem the domain cost is the sum of the lengths of the vehicle routes). Several economic models for bounded rational agents have been proposed [305], [269], but these approaches have been more descriptive rather than prescriptive. Recently, Larson and Sandholm integrated deliberation actions into agent strategies and analysed game theoretically bounded rational agents in a 2-agent bargaining game [173] and in auctions [171], [172].

The negotiation model defined in the context of the NPI framework is an approach inspired by non-cooperative models. Bounded rationality stems from boundaries, limitations and constraints that are intrinsic to the CSP-based decision making model that our agents rely upon.

Among various existing heuristic approaches, the work of Faratin is particularly relevant to the work. In the proposed service-oriented negotiation context [82], one agent (the client) requires a service to be performed on its behalf by some other agent (the server). Negotiation involves determining a contract under certain terms and conditions. The negotiation may be iterative in that several rounds of offers and counter offers will occur before an agreement is reached or the negotiation is terminated. More precisely, a formal model of a negotiating agent reasoning component is proposed. The focus is on the processes of generating an initial offer, of evaluating incoming proposals, and of generating counter proposals. Agents can also make trade-offs between the various issues over which they are negotiating. The model specifies the key structures and processes involved in terms of tactics and strategies. The *wrapper* component and the developed deliberation mechanisms have been applied to business process management [151] and telecommunication network management [83]. The reasoning model proposed by Faratin is in some way close to the CSP approach proposed in this book. Relaxing constraint in terms of QoS requirements or connectivity is equivalent to making concession. Choosing another variable value in the set of equivalent intra-domain routes for iterative proposals is similar to making trade-offs. However, while in the NPI framework, agents strategies and coordination mechanisms are directly integrated within resource allocation and control processes for managing communication networks, by automatically taking into account the specific settings, requirements and constraints of the multi-provider environment, Faratin proposes a more abstract (i.e., application independent framework) that have been evaluated in different areas (such as business process management, telecom service management).

2.5   Distributed Constraint Satisfaction Approach

Constraint satisfaction is an efficient and extensively used AI paradigm that involves finding values for problem variables subject to restrictions (constraints)
on which combinations of values are acceptable. The success and the increasing
application of this paradigm in various domains mainly derived by the fact that
many combinatorial problems can be expressed in a natural way as a *Constraint
Satisfaction Problem* and can subsequently be solved by applying powerful CSP
techniques.

A finite discrete CSP is defined by:

- a finite set of *variables V* whose values are taken from

- a finite, discrete *domain D* and

- a finite set of *constraints C* restricting the values that variables can simultaneously take.

The *arity* of a CSP is defined by the number of variables involved in a constraint
(more formal CSP models can be found in [192], [169] and   [65]). This basic
definition has been extended in many ways allowing dynamic sets of variables,
dynamic, continuous or infinite domains, and constraints of various arity. Solving
a CSP means finding an assignment of values to all variables, from their separate
domain, such that all constraints are satisfied.

A DCSP is a constraint satisfaction problem in which the variables, domains
and constraints are distributed among multiple distinct autonomous processing
entities also called *agents* [368]. Each agent has one or multiple variables and
tries to determine its/their value/s. However, there exist intra- and inter-agent
constraints and the value assignment must satisfy all these constraints. Figure 2.7
shows a simple example of DSCP where three distinct agents, A, B, and C control
three distinct variables, respectively v_A, v_B and v_C. The domains for each variable
are $D_A = \{5, 10, 15\}$, $D_B = \{8, 12, 20\}$ and $D_C = \{10, 15, 25\}$ and the existing
constraints are $C(v_A, v_C) = (v_A \leq v_C)$ and $C(v_B, v_C) = (v_B > v_C)$. There exist
several possible value assignments to variables v_A, v_B and v_C that solve the DCSP.
For instance, the assignment $v_A = 5$, $v_B = 15$ and $v_C = 20$ is a solution since all
existing constraints are satisfied. There is no possible solution for $v_B = 8$ or for
$v_C = 25$.

In the DAI area and in related fields, many problems that require finding of a
consistent combination of agent actions can be formalised as DCSPs. Among various works, some significant examples include multi-agent truth maintenance [134],
distributed interpretation problem [199], distributed job-shop scheduling [310],
in particular, nurse time-tabling task [183], resource allocation approach in distributed communication network [57], [372] and frequency assignment for cellular
mobile systems [371]. An early overview of DCSP models and algorithms is given
in [187]. Luo, Hendry and Buchanan distinguish three main classes of DCSP solving approaches: *variable-based* approaches, in which every agent is responsible for

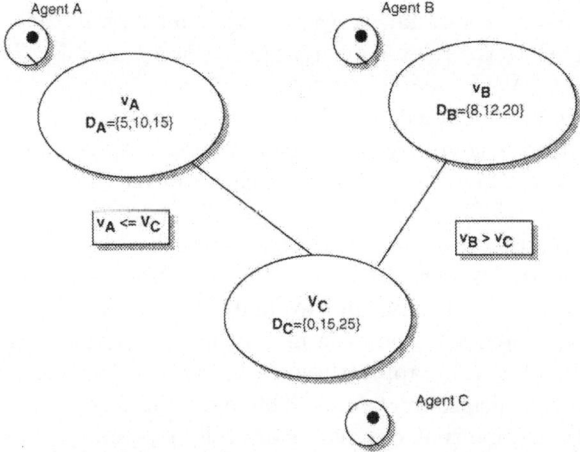

Figure 2.7: *In this simple example of DCSP there are three distinct variables, v_A, v_B and v_C, that agents A, B and C respectively control. The domains for each variable are $D_A = \{5, 10, 15\}$, $D_B = \{8, 12, 20\}$ and $D_C = \{10, 15, 25\}$ and the existing constraints are $C(v_A, v_C) = (v_A \leq v_C)$ and $C(v_B, v_C) = (v_B > v_C)$.*

a subset of variables, *domain-based* approaches, in which every agent is responsible for a subset of values of a unique variable, and *function-based* approaches in which costly computations in centralised CSPs are distributed in order to speed them up. Following this classification, the NPI approach is variable-based. The most trivial algorithms for solving DCSPs are the *centralised method* and the *synchronous backtracking* approach. In the first case, a leader agent gathers all information about the problems, their domains and their constraints and then solves the CSP alone using classic centralised constraint satisfaction algorithms. This approach is very inefficient since there are costs associated to gathering information, there are unused resources (i.e., agents sitting idle) and there is no confidentiality. The synchronous algorithm assumes that agents do agree on an instantiation order. The first agent generates a partial solution that is submitted to the second agent, which generates an extension to the partial received solution and sends it to the third agent and so on. If the solution cannot be extended then a backtracking message is sent to the previous agent. With this approach there are still costs to decide and to transmit the instantiation order; furthermore, only one agent is active at time. Several *asynchronous search* algorithms that allow agents to work in parallel and asynchronously have also been developed. For instance, in the asynchronous backtracking (ABT) algorithm agents work in parallel and asynchronously. The ABT algorithm applies to binary and directed constraints [368] (i.e., for every binary constraint between any two variables only one of the two agents is responsible for evaluating the constraint). Every agent instantiates its variable v_i concurrently and send v_i assigned value to the agents which own variables constrained by v_i

value. When a constraint evaluating agent detects a constraint violation, a *nogood* message is sent back to the related agents causing backtracking. The asynchronous weak commitment (AWT) search proposes a variation of the ABT algorithm: a partial solution is not modified, but completely abandoned after a failure [372]. The AWT algorithm is complete if all nogood messages are maintained and it is about 10 times faster than the ABT approach. However, the explosion of nogood messages is the most difficult part to control. To coordinate the different forms of asynchronous interactions, the algorithms establish a static or a dynamic order among agents that determines the cooperation patterns between agents. Nishibe, Yokoo and Ishida [213] discuss and evaluate asynchronous backtracking with different ordering schemes: value ordering, variable ordering and value/variable ordering. In particular, they apply these techniques to the communication path assignment in communication networks. The main difference with the NPI framework is that they assume that the agents are fully cooperative since they control different resources of the same distributed network, i.e., unique administrative domain.

More exhaustive surveys and descriptions of algorithms and methodologies for solving DCSPs, including the ABT and the AWS approaches, can be found in [374], [369], [373], [118].

2.5.1   DCSP-based Coordination and Agents: Related Work

Several works have deployed DCSP oriented formalism and techniques in order to enforce coordination in multi-agent systems and efficiently model and solve constraint-directed negotiation approaches.

One of the earliest approaches that modelled the process of negotiation as a distributed constraint satisfaction problem is the *constraint-directed negotiation* paradigm [286]. Here, negotiation for resolving conflicts, i.e., coordination, in resource allocation is modelled as constraint relaxation and constraints are used for evaluation of existing alternatives as well as for creating new ones. Three main constraint-directed negotiation algorithms are proposed and experimentally validated. A centralised mediator clusters several announcements and bids from multiple agents into atomic contracts (i.e., bids are grouped into *cascades*). Experimental results are discussed for the real world problem of workstation requirements within an engineering organisation: the resources are workstations that each group within the organisation uses. When projects change the groups requirements vary and therefore the initial allocation of resources have to be adapted. This kind of coordination is not always possible in scenarios where the centralisation of information may not be feasible. A fundamental limitation of the constraint-directed negotiation approach is that only constraints that are qualitative in nature are considered. In the NPI framework, utility functions have to be able to model both qualitative and quantitative constraints. Furthermore, specific search operators used for converging a global solution that accommodates all negotiation participants, such as the *relaxation* and the *reconfiguration* operators, are directly related to the specific domain they apply to and no formal and general re-usable models are given. However, the approach presented by Sathi and Fox in [286] considers several aspects that are relevant and common to this work: the CSP-based decision making model, the direct mapping between agents preferences and constraints, and the use of negotiation timeouts.

Another distributed constraint satisfaction based approach for coordination in multi-agent systems has been proposed by Liu and Sycara in [152]. The main idea in their framework is to partition the problem constraints into constraint types. Responsibility for enforcing constraints of a particular type is given to *specialist* agents that coordinate to iteratively change the instantiation of variables under their control according to their specialised perspective. Agents converge to the final solution through incremental local revisions of an initial, possible inconsistent, instantiation of all variables. This approach is valid in the domain of job shop scheduling.

A centralised approach integrating constraint techniques within a multi-agent system for developing a decision support system in production flow control (PFC) has been presented in [14]. The relevant part of this work to this work is in the use of constraints for expressing relations between different entities and restrictions over resources and structures. The idea of using constraints propagation in order to coordinate several interacting parts is also close to the NPI approach. However, in our case there is no central constraint-solver as in the PFC system.

In [98], Freuder and Wallace model "content-focused matchmaking" as a constraint satisfaction problem in which negotiation techniques are used. The constraint solver interacts with a human providing partial solutions (suggestions), based on partial knowledge, and gaining further information about the problem from the humans evaluation of the suggestions. Several strategies that might be useful in this context are proposed and evaluated experimentally. We believe that intelligent and automated matchmaking could be easily integrated within the NPI approach from the end user perspective, for instance, whenever a combination of Telecom services is needed.

Integrated and automated management of supply chains by means of negotiating agents deploying the CSP paradigm have recently been proposed by Sun et al. in [316]. Buyers and sellers make and/or evaluate bids by considering possible terms and conditions as CSP variables and their domains. An acceptable offer is a set of variable assignments that satisfy all constraints. Although, the decision making process of agents in this book has been modelled in a very similar way to the paradigm proposed in [316] there is a fundamental difference. Constraint satisfaction techniques are used in the supply chain framework to model a single agent decision making process: there is no propagation of constraints and no dynamic coordination between distributed entities. In this context, constraint satisfaction techniques are used at two different levels: at the intra-agent level for modelling the decision making process and at the inter-agent level for enforcing coordination by means of specific consistency techniques.

Finally, a very recent framework proposes an experimental system of e-Negotiation Agents (eNAs) [161] where the negotiation problem is modelled as a constraint satisfaction problem and the negotiation process as constraint-based reasoning. Agents use the branch and bound search with the support of constraint propagation to find instantiations that satisfy the constraints of the party. However, it is not clear from the accessible results what is the amount of private information that has to be disclosed in order to achieve consistent solutions. Therefore, it is not evident to understand if self-interests prevail against cooperativeness.

Chapter 3

Definitions and Multi-Provider Problem Formalisation

> *One benefit of a modelling effort,*
> *sometimes obtained unintentionally,*
> *is that it forces an improved understanding*
> *of what is really happening in the system*
> *being modelled.*
>
> – P. BRATLEY, B. FOX, L. SCHRAGE, *A Guide to Simulation, 1987*

In order to formally define the multi-provider service set up process, a model of the underlying communication networks, the services offered, and the end-to-end communication demands need to be specified. These models represent a simplified and abstract view of real concrete networks and communication services in a way that make it possible to effectively solve the MuSS problem. Given the deployment of techniques and methods from different research and technology fields as well as their integration within a unified framework for solving problems in the networking field, a preliminary step to avoid ambiguity is the definition of terms and concepts (see Section 3.1).

Then, the central part of the chapter (i.e., Section 3.2) goes on to formalise:

- The main assumptions made when modelling the MuSS problem, with the twofold purpose of (1) abstracting from the low level technical details of networks implementations, and (2) simplifying the description of the proposed techniques.

- The resource model underlying every distinct network, including single network elements such as nodes and links.

- The abstract model of a set of several interconnected networks in which each network is considered a single node. This aggregated, abstract information represents the global view that every provider may have of the overall multi-provider scenario.

- The service model consists of four main classes of services that differ for the QoS level they offer. This model is assumed to be valid for all the providers.

- A service demand whilst adhering to certain QoS properties expresses the requirement on the network for transferring data between specific nodes. The definition of a formal model makes it possible to identify which specific QoS properties characterise a service demand.

The formalisation of service and network components introduces the next step that this chapter discusses: the formalisation of the MuSS problem by means of a distributed constraint satisfaction-based formalism (see Section 3.3). The central idea is that variables (local network resources) and constraints on which combinations of values are acceptable for these variables are distributed over a set of agents acting on behalf of distinct providers.

The final part of the chapter (Section 3.4) concentrates on the network resource abstraction techniques that have been applied in this work in order to reduce the complexity of intra-domain tasks such as routing, resource allocation and pricing. The Blocking Island (BI) paradigm, previously developed by Frei [93], is illustrated and the main concepts that have been refined and re-used in this book are discussed.

3.1   Fundamentals Terms and Concepts from Networking

One of the unfortunate side effects of the separation of work between different research groups such us the MAS, the DCSP and the networking communities is confusion over terminology. A typical example is the word *agent* that has been used with different meanings[1]. In order to facilitate the reading of this volume, some of the main concepts and terms that need to be uniquely defined to formalise and understand the NPI approach are introduced in this section. Many other definitions are given whenever specific topics or issues are raised[2].

---

[1] The definition of agent adopted in this framework has strong DAI origins and is given in Section 2.4.1. An overview of existing definitions of agents in a network management context is given in [197].

[2] For more details on terms and definitions relevant to communication and data networks several on-line glossaries can be consulted (i.e., the Telemanagement Forum Glossary can be

Definition 3.1.1 (Customer). This term refers to companies or private end users that make use of telecommunications services provided by a *service provider* (see below). In other words, the customer is the ultimate buyer of a network service. In this volume, the term 'customer' is equivalent to the term 'end customer'.

Definition 3.1.2 (End User). An end user (or simply a user) is in the domain of the customer and he/she is interested in using communications and data services, e.g., Telecom, Internet/Intranet, mobile, wireless, etc.

A company that outsources network resources from external providers, for instance, represents the customer, while the people in the company making use of the network based services are end users.

Definition 3.1.3 (Interworking). This term refers to the set of operations and processes involved in multi-provider interactions for the allocation of service demands spanning distinct networks.

Definition 3.1.4 (Network Operator). An organisation that operates a communication network, network or data services capabilities, acting basically as a wholesaler. A network operator owns and directly controls his networks resources and may also act as a service provider (see Definition 3.1.10). In this context, the expression 'network operator' is equivalent to the expression 'network provider'.

Definition 3.1.5 (Quality of Service). Quality of Service is a very broad concept which comprises many performance aspects as well as numerous measures. There are many different definitions for QoS depending, for instance, upon whether they are defined from the viewpoint of customer, service or network provider (for more formal definitions see [141], [223]). In this work, the QoS is given by the selection of a combination of QoS parameters and the specification of their target values. A *QoS parameter* [142] is a variable that characterises certain aspects of QoS, such as, for example, the required bandwidth, the maximal end-to-end delay, the mean access time, etc.

Definition 3.1.6 (Service). A communication service is a set of independent functions that are an integral part of one or more business processes. This functional set consists of hardware and software components as well as the underlying communication medium. The customer sees both as an amalgamated unit.

Definition 3.1.7 (Service demand). A service demand expresses the requirement on the network for transferring data between specific nodes whilst adhering to specific QoS properties.

Definition 3.1.8 (Service Access Point). A service access point is a logical element located on the interface between the customer domain and the service provider domain and it represents the point to which a service is delivered.

---

found at the address: http://www.tmforum.org. For definitions of the most current IT-related concepts a good glossary can be found at http://WhatIs.techtarget.com/.

Definition 3.1.9 (Service Level Agreements). Service Level Agreements (SLAs) can be considered as part of a contract between a customer and a provider or between distinct providers. A SLA represents a negotiated agreement which defines the service to be provided and the set of metrics to be used to measure the level of service provided. Such a service level might include network performance metrics (i.e., QoS, installation completion, time metrics, availability, peak rate, etc.) and define trouble reporting and the general responsibilities of both parties.

Definition 3.1.10 (Service Provider). This term refers to companies who provide communication and/or data services as a business. Service providers may operate networks (i.e., they may own and directly control network resources) or they may simply integrate the service of other network providers (who operate networks) in order to deliver a service to their customers.

Definition 3.1.11 (Unicast). This term indicates a communication between a single sender and a single receiver over a network in contradistinction to *multi-cast* that refers to a communication between a single sender and multiple receivers. The term *point-to-point communication* is equivalent in meaning to unicast.

3.2 Problem Modelling

The allocation of service demands spanning several network domains involves a set of activities needed to determine a specific route that satisfies connectivity and QoS constraints and which all peer providers agree upon. This is a very complex task for several reasons:

- There are distinct entities involved (final customers, service and network providers, etc.) that need to coordinate their actions despite possible conflicting goals, policies and objectives. This requires finding a trade-off between the providers profit optimisation and end users satisfaction (or utility). In economic terms, this means that the social welfare should be maximised in order to accommodate both customers and providers requirements. The social welfare is considered the benefit to the society measured by the sum of providers and end users benefit (see Definition 5.1.11).

- The demand routing process must take into account QoS requirements and the existence of different underlying network technologies that can or cannot support such requirements. In general, such a kind of routing problem is NP-complete [342].

- Network resources and information are distributed. This generates a flow of messages exchanged for recovering data about boundary network elements and for maintaining management information databases mapping the current global state (see Definition 4.3.4) of multi-provider networks (see Section 4.3.2).

In the following, a formal way of modelling a simplified version of the MuSS problem is presented. This description abstracts from technological details of underlying networks and identifies a level that aims to provide mechanisms for improving peer-to-peer interactions in different possible scenarios. Later in this volume (see Section 8.5), the feasibility of this solution is discussed with respect to several existing network technologies.

3.2.1 Assumptions

In order to abstract from problems connected to the low level technical details of networks implementations and simplify the description of the techniques proposed in this book, several assumptions have been made.

Assumption 3.2.1. All network providers in the multi-provider scenario are uniquely identified.

Every provider has a unique name called *provider-id*, and every software entity acting on behalf of the provider has a unique agent identifier indicated as *agent-provider-id*. The *agent-provider-id* is built by adding the suffix "NPA" to the *provider-id*, where the acronym NPA stands for *Network Provider Agent*. For instance, given the provider named 'Ultimate' its representative agent is called 'UltimateNPA'.

Assumption 3.2.2. All network providers periodically communicate to maintain a global aggregated view of all existing peer providers in the scenario.

This means that all network providers in the multi-provider scenario know each other. In order to maintain a global and aggregated (partial) view of all peer operators, provider agents periodically exchange messages between each other. For this purpose, a link-state protocol [231] has been implemented, where update messages only contains a minimal amount of information necessary to identify the provider and advertise average performance in terms of traversal cost and delay (see Section 4.3.2).

Assumption 3.2.3. Communication links are bidirectional.

This assumption indicates that service demands can be allocated over the links regardless of the data flow direction, and the proportion of information flowing in one direction is not fixed. This assumption allows the network to be modelled as a *non-directed* graph (see Section 3.2.3).

Assumption 3.2.4. All service demands are unicast.

As specified in Definition 3.1.11, a unicast service demand corresponds to a communication between a single sender and a single receiver over the network, i.e., a point-to-point communication.

Assumption 3.2.5. The only service demand QoS parameters taken into account are the bottleneck bandwidth and the end-to-end delay.

For most applications, particularly real-time ones, the combination of the bottleneck bandwidth, i.e., minimal bandwidth required, and end-to-end delay represent the most important QoS requirements [342]. The bandwidth is typically measured in bits per second and once allocated to a given demand it cannot be used by another demand, i.e., bandwidth is a non-sharable resource. The end-to-end delay indicates the delay that traffic experiences through the network. This parameter is influenced by the elements that constitute the network, the traffic that goes through these elements and the way the network is operated. The total end-to-end delay is usually decomposed into four elements:

- The time required to transmit a certain amount of data, i.e., the amount of data divided by the transmission speed. For example, for a 10000-bit packet and a transmission speed of 1 Mbps, this component is equal to 10 ms.

- The signal propagation time computed as the distance from source to destination divided by the speed of the electrical or optical signal. Propagation time for an electrical or optical signal is between 3.3 and 5 μs/km.

- The queueing delay (or buffering delay) in the intermediate switches and/or routers.

- The processing time required by the switches and/or routers to forward data.

In this context, the computation of the end-to-end delay (see Section 3.2.3) is simplified by considering network nodes with an infinite buffering capacity and infinite computation resources, i.e., the transmission delay, the queueing delay and the processing time are negligible. Concerning traffic characteristics such us mean inter-arrival time, burstiness, etc., the main assumptions valid in different simulated scenarios are described in Chapter 7.

Assumption 3.2.6. Demands must be allocated right away, but a specific end time is given.

This means that every service demand is characterised by a specific duration, which is the time interval during which network resources have to be allocated. By considering demands that can start at any future instant the complexity would strongly increase. To deal with this issue would require to make plans for the future based on forecast about future incoming traffic. This is out of the scope of this work.

Assumption 3.2.7. Service demands have constant QoS requirements for their whole lifetime.

Since for Assumption 3.2.5 the considered QoS requirements are bandwidth and end-to-end delay, Assumption 3.2.7 implies that demands are constant bit rate (CBR) demands with a constant end-to-end delay constraint.

In Section 8.2, the relaxation of some of these restrictions and the subsequent main implications on the NPI approach are discussed.

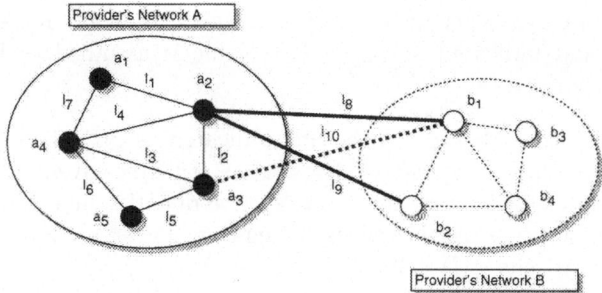

Figure 3.1: *Intra-domain links have both end-points directly controlled by a unique provider. Inter-domain links have one end-point directly or partially controlled by another provider.*

3.2.2 Single Provider Network Model

A communication network consists of nodes (i.e., routers, switches, computers, phone terminals, etc.) interconnected by transmission links (i.e., communication media such as copper wires, optical fibers, microwave links, and satellite transmissions). In this book, each network is therefore modelled by a connected *domain graph* $\mathcal{G} = (\mathcal{N}, \mathcal{L})$, a non-directed multi-graph without loops[3], where the nodes $\mathcal{N}$ correspond to the network nodes or sub-networks, and the edges $\mathcal{L}$ represent the set of bidirectional communication channels existing between the network nodes. The two network nodes that a specific link l interconnects are the *end-points* of l and they are assumed to always be distinct. The main distinction valid in this framework is between intra- and inter-domain links.

Definition 3.2.8 (Intra-domain Link). A link that has both end-points entirely controlled and owned by the same provider is an intra-domain link.

In the example depicted in Figure 3.1, for instance, l_1, l_2, l_3, l_4, l_5, l_6 and l_7 are intra-domain links under the direct control of the provider A.

Definition 3.2.9 (Inter-domain Link). A link that has one end-point totally or partially controlled by another provider is an inter-domain link.

Inter-domain links are directly controlled by either one or the other provider owning the nodes that the links are interconnecting. In the example depicted in Figure 3.1, for instance, l_8 and l_9 are inter-domain links controlled by the provider A, while l_{10} is an inter-domain link controlled by the provider B.

Bandwidth and end-to-end delay are the main links QoS properties that are considered in the NPI framework (see Assumption 3.2.5). More precisely, every link $l \in \mathcal{L}$ (both inter- and intra-domain) is characterised by the following set of parameters:

---

[3] Without loops means that there are no edges whose end-points are the same nodes.

- The total link capacity in terms of bandwidth (typically measured in [bits/ second]) and indicated as $\beta_c(l)$. This value is assumed to be a constant positive quantity.

- The available bandwidth on the link indicated as $\beta(l)$. This parameter expresses the amount of bandwidth that is still free given all the demands currently allocated. This value is a positive quantity depending on the network state, i.e., allocated services. When the allocated bandwidth equals the total link capacity $\beta(l) = 0$.

- The end-to-end delay $e(l)$ computed as the distance metric from source to destination divided by the speed of the electrical or optical signal. This value is assumed to be constant.

- The cost $c(l)$ of the link that is a variable value, since it is considered to quantify the opportunity cost of reserving and allocating bandwidth capacity on the link. This parameter estimates the cost of deploying network resources from the provider perspective.

Moreover, every provider dynamically computes the price of a link that is used to allocate a specific service demand. This corresponds to the amount of money that an end user is asked to pay for using resources on a specific link. Dynamic link costs and price computations are extensively considered in Chapter 5, where a discrete cost function and a specific pricing scheme are proposed and discussed.

Network nodes, typically gateways, routers and switches, are classified as *internal*, *boundary* and/or *shared* nodes. Considering for instance the DS framework, internal nodes correspond to core routers, while boundary and shared nodes correspond to edge routers.

Definition 3.2.10 (Internal Node). A network node belonging to a specific provider network I is an internal node of I, if all the links directly connected to it are intra-domain links belonging to I.

Definition 3.2.11 (Boundary Node). A network node belonging to a specific provider network I is a boundary node of I if at least one of the links directly connected to it is an inter-domain link[4].

Definition 3.2.12 (Shared Node). A network node is a shared node between different (two or more) provider networks if there exist at least one inter-domain link per provider network directly connected to the node.

In other words, considering two distinct networks I and J, a network node $n \in I$ is a shared node if there exist at least an inter-domain link l_x belonging to I and a distinct inter-domain link l_y belonging to J directly connected to n. In the

---

[4]Notice that the inter-domain link directly attached to a boundary node can also belong to another domain.

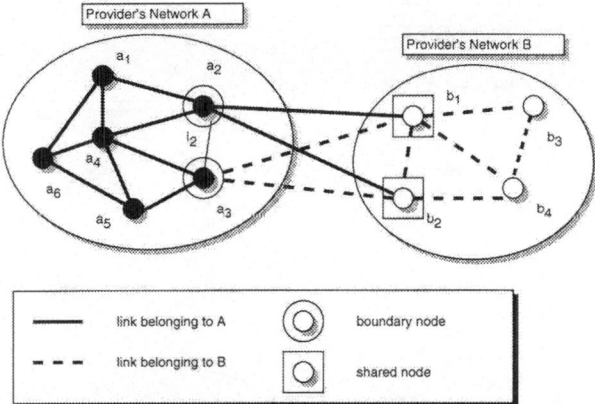

Figure 3.2: *Network nodes are classified in three distinct categories: internal, boundary and shared nodes. Shared nodes represent the common resources that providers share in order to interconnect their domains. In real networks, these nodes are typically gateways.*

example represented in Figure 3.2, it is possible to visualise the different types of nodes. Considering network nodes with an infinite buffering capacity and infinite computation resources nodes costs and delay are assumed to be negligible.

In this context, some basic properties hold.

Proposition 3.2.13. *A shared node is always a boundary node.*

PROOF: This is obvious from the definitions of boundary node and shared node (see Definitions 3.2.11 and 3.2.12). □

In the example depicted in Figure 3.2, the shared nodes b_1 and $b_2 \in B$ are also boundary nodes.

Proposition 3.2.14. *A boundary node is not necessarily a shared node.*

PROOF: A boundary node which is endpoint only for links belonging to I cannot be a shared node by definition (see Definition 3.2.12). □

In the example depicted in Figure 3.2, the boundary nodes a_2 and a_3 in network A are not shared nodes since they are directly connected only to links belonging to A.

3.2.3 Multi-Provider Network Model

When considering a set of distinct interconnected networks, it is possible to work at a higher level of abstraction. This level is referred to as the multi-provider level at which every network belonging to a specific operator can be represented as a unique abstract (or global) node.

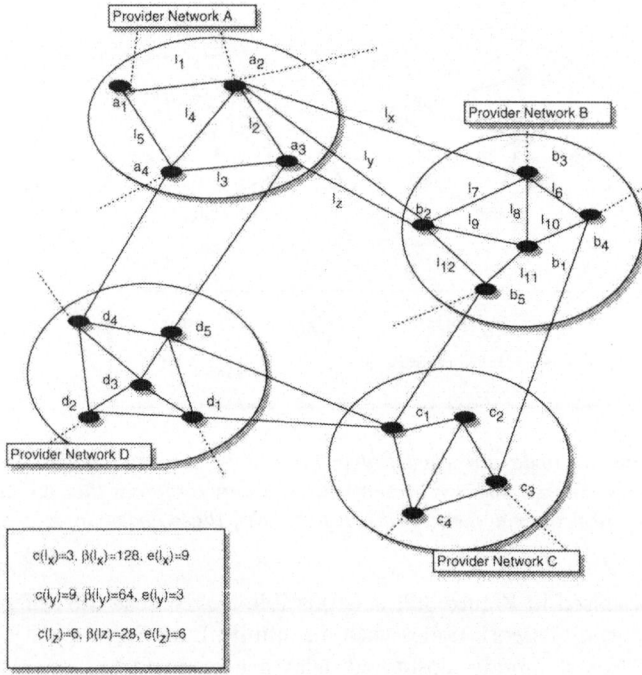

Figure 3.3: *Four distinct provider networks are interconnected to form a multi-provider network scenario. For more clarity the name of the links are omitted in networks C and D.*

Definition 3.2.15 (Provider Network). A provider network is a communication network consisting of all the nodes and links directly controlled and owned by a specific provider.

Considering several interconnected provider networks, the concept of abstract path is defined as follows.

Definition 3.2.16 (Abstract Path). An abstract path $\mathcal{A}$ between two network nodes $i \in \mathcal{N}_I$ and $j \in \mathcal{N}_J$, with $\mathcal{G}_I = (\mathcal{N}_I, \mathcal{L}_I)$ and $\mathcal{G}_J = (\mathcal{N}_J, \mathcal{L}_J)$ representing two distinct operators I and J, is an ordered list of provider networks between I and J, including the latter two domains.

The networks belonging to an abstract path are ordered linearly from the source to the destination network. This means that, given a certain abstract path $\mathcal{A}$, a specific network J precedes a network I when, going from the source to the destination networks of $\mathcal{A}$, the network J is encountered before network I. Analogously, given a certain abstract path $\mathcal{A}$, a given network J succeeds a network I when, going from the source to the destination networks of $\mathcal{A}$, the network J is encountered after network I. In Figure 3.3, for instance, four provider networks are represented, namely network providers A, B, C and D. In this example, the set of possible abstract path between $a_1 \in \mathcal{N}_A$ and $b_4 \in \mathcal{N}_B$ is the set $S = \{A - B, A - D - C - B\}$. Along the path $\mathcal{A} = A - B$ network A precedes network B and network B succeeds network A, there are no networks preceding A and no networks succeeding B.

Definition 3.2.17 (Initiator Provider). Given an abstract path $\mathcal{A}$ between two networks I and J, the provider controlling the source network I is called the *initiator* provider.

This definition directly applies also to the network the initiator provider is controlling. Therefore, in this volume the expressions 'initiator network' and 'initiator provider' are both valid.

Definition 3.2.18 (Predecessor Provider). Given an abstract path $\mathcal{A}$ between two networks and a specific provider $X \in \mathcal{A}$, the provider P controlling the network preceding X along $\mathcal{A}$ is called the *predecessor* provider of X.

Definition 3.2.19 (Successor Provider). Given an abstract path $\mathcal{A}$ and a specific provider $X \in \mathcal{A}$, the provider P controlling the network succeeding X along $\mathcal{A}$ is called the *successor* provider of X.

Considering the example in Figure 3.3 and the abstract path $\mathcal{A} = A - D - C - B$, A is the *initiator* provider of $\mathcal{A}$ and the *predecessor* provider of the domain D along $\mathcal{A}$. Provider D is therefore the *successor* of provider A and the *predecessor* of provider C and so on.

Definition 3.2.20 (Global End-to-end Route). Once specific intra-domain links are selected along a given abstract path $\mathcal{A}$, the whole sequence of links interconnecting the source to the destination nodes is called the *global end-to-end route*.

In the example depicted in Figure 3.3, a possible global end-to-end route between $a_1 \in \mathcal{N}_A$ and $b_4 \in \mathcal{N}_B$, along the selected abstract path $\mathcal{A} = A - B$, is given by the sequence of links l_1-l_x-l_6.

Definition 3.2.21 (Provider Set). A provider set $\mathcal{P}$ is an abstract simple graph consisting of global nodes and global links abstracting the interconnection of different provider networks. A *global node* abstracts all the resources of a specific provider. A *global link* between two distinct global nodes clusters all inter-domain links between the two corresponding provider networks.

In Figure 3.4, the provider set corresponding to the network scenario depicted in Figure 3.3 is shown. Every global node of a given provider set $\mathcal{P}$ represents a distinct provider network and is characterised by a *node traversal cost* and a *node traversal delay*. More precisely, these parameters are defined as follows:

Definition 3.2.22 (Node Traversal Cost). For a given provider set $\mathcal{P}$, every network node n belonging to $\mathcal{P}$ is characterised by the node traversal cost $c(n)$ which is an estimated cost of traversing a provider network.

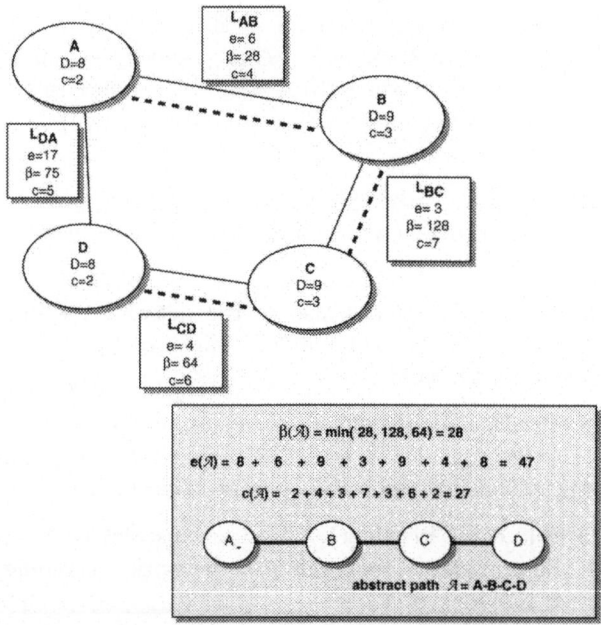

Figure 3.4: *The network model at the highest level of abstraction: the provider set of the scenario depicted in Figure 3.3. For every global node the node traversal delay $\mathcal{D}$ and the node traversal cost c are displayed. For every global link the estimated cost c, the end-to-end delay e and the available bandwidth β are displayed. The estimated end-to-end delay (cost) of the abstract path $\mathcal{A} = A - B - C$ is computed as the sum of the traversal delays (costs) of all individual global nodes forming the path and the delays (costs) of all the global links interconnecting them. The available bandwidth of the abstract path is the minimum of the available bandwidth on the interconnected global links forming it.*

Definition 3.2.23 (Node Traversal Delay). For a given provider set $\mathcal{P}$, every network node n belonging to $\mathcal{P}$ is characterised by the node traversal delay $\mathcal{D}(n)$ which is an estimated delay of traversing the provider network n.

$c(n)$ and $\mathcal{D}(n)$ are metrics determined either by considering average values experienced for past service demand allocations or by considering values explicitly advertised by peer network providers through the use of a link-state protocol[5].

Definition 3.2.24 (Global Link). A *global link* L between two distinct global nodes clusters all inter-domain links between the two corresponding provider networks. Some of these links can be directly controlled by one provider and some others by its neighbours.

---

[5]The traversal cost is an estimated value used by every provider in order to decide which abstract path (or in which order several abstract paths) will be explored for starting the end-to-end demand allocation process. The effective global cost for crossing a provider network depends on which specific route has been negotiated and selected (see Chapter 5).

For instance, in the example depicted in Figure 3.4, the global link L_{AB} corresponds to the physical links l_x, l_y, l_z, with, for instance, l_x and l_y controlled by provider A and l_z controlled by provider B. Every global link is characterised by an estimated cost $c(L)$, an estimated end-to-end delay $e(L)$ and its available bandwidth $\beta(L)$.

Definition 3.2.25 (Cost of a Global Link). The estimated cost of a global link L is computed as the average value obtained by considering all the corresponding physical links:

$$c(L) = \sum_{i=1}^{L} c(l_i)/|L|.$$

where $c(l_i)$ is the cost of every physical link $l_i \in L$ and $|L|$ is the cardinality of set L indicating the total number of physical links clustered by the global link L.

In the example depicted in Figure 3.4, the estimated cost of the global link L_{AB} is $c(L_{AB}) = (c(l_x) + c(l_y) + c(l_z))/3 = 6$.

Definition 3.2.26 (End-to-end Delay of a Global Link). The estimated end-to-end delay of a global link L is computed as the average delay of the corresponding physical links delays:

$$e(L) = \sum_{i=1}^{L} e(l_i)/|L|.$$

where $e(l_i)$ is the end-to-end delay of every physical link $l_i \in L$ and $|L|$ indicates the total number of physical links clustered by the global link L.

In the example depicted in Figure 3.4, the estimated delay of the global link L_{AB} is $e(L_{AB}) = (e(l_x) + e(l_y) + e(l_z))/3 = 6$.

Definition 3.2.27 (Bandwidth of a Global Link). The bandwidth $\beta(L)$ of a global link is given by the minimal amount of available bandwidth on the corresponding physical links $l_i \in L$:

$$\beta(L) = \min(\beta(l_1), \beta(l_2), ..., \beta(l_N)).$$

where $\beta(l_i)$ the available bandwidth on every physical connection $l_i \in L$.

The available bandwidth corresponds to the maximum amount of bandwidth that can still be allocated on a link. It can be computed as the difference between the total nominal capacity of a link and the currently deployed bandwidth, given the set of demands already allocated in the network. Considering the example of Figure 3.4, L_{AB} has $\beta(L_{AB}) = \min(\beta(l_x), \beta(l_y), \beta(l_z)) = 28$.

Considering a provider set, an abstract path (see Definition 3.2.16) between two network nodes i and j belonging two different provider networks can also be seen as an ordered list of global nodes between the two global nodes clustering respectively i and j, including all the abstract links between the global nodes. For every abstract path $\mathcal{A}$, three main parameters are considered: the estimated end-to-end delay $e(\mathcal{A})$, the available bandwidth $\beta(\mathcal{A})$ and the estimated global cost $c(\mathcal{A})$. These parameters are computed as follows:

$$e(\mathcal{A}) = \sum_{i=1}^{L} e(L_i) + \sum_{j=1}^{N} \mathcal{D}(N_j) \tag{3.1}$$

$$\beta(\mathcal{A}) = \min_{i \in \{1,\dots,L\}} (\beta(L_i)) \tag{3.2}$$

$$c(\mathcal{A}) = \sum_{i=1}^{L} c(L_i) + \sum_{j=1}^{N} c(N_j) \tag{3.3}$$

Notice that $L = |N| - 1$ is the total number of global links interconnecting the global nodes N forming the abstract path $\mathcal{A}$. For a better understanding, let us consider the example depicted in Figure 3.4. The abstract path $\mathcal{A}$ formed by the four distinct networks A, B, C and D, for instance, has the following characterising parameters:

- A global estimated end-to-end delay $e(\mathcal{A}) = (e(L_{AB}) + e(L_{BC}) + e(L_{CD})) + (\mathcal{D}(A) + \mathcal{D}(B) + \mathcal{D}(C) + \mathcal{D}(D)) = 47$

- $\beta(\mathcal{A}) = \min(28, 128, 64) = 28$ available bandwidth.

- An estimated global cost $c(\mathcal{A}) = c(A) + c(L_{AB}) + c(B) + c(L_{BC}) + c(C) + c(L_{CD}) + c(D) = 27$.

3.2.4 Service Model

The set of services offered by a network provider is called *service model*. In the NPI framework, the same service model is assumed to be valid for all providers and it consists of four main classes of service that differ for the QoS level they offer. The definition and the adoption of a common service model implies the use of a common ontology[6] that facilitates the coordination of different providers by encompassing heterogeneity of network components and intra-domain information models. The common and discrete model adopted in this context has been defined by considering the current service level differentiation that the IETF community proposes for the Differentiated Framework [23] in the Internet. The *Gold* level corresponds to stringent QoS requirements, e.g., virtual leased lines. The *Silver* level to an amount of capacity to experience a good[7] quality and the *Bronze* level to just enough capacity to carry the traffic on the average. Finally, the *Premium* level symbolises the classical best-effort traffic. In this context, a specific QoS

---

[6] An ontology provides a vocabulary for representing and communicating knowledge, including a set of relationships and properties that are valid for the elements identified by that vocabulary.

[7] Goodness is relative to the specific service.

class is defined by considering *bandwidth* and *end-to-end delay* characteristics. This generic model could be extended in order to take into account other parameters, such as loss ratio, reliability, etc., depending on the specific underlying network technology. This is better discussed in Section 8.3.

Given two distinct provider networks I and J and their respective domain graphs, $\mathcal{G}_I = (\mathcal{N}_I, \mathcal{L}_I)$ and $\mathcal{G}_J = (\mathcal{N}_J, \mathcal{L}_J)$, the specified service model and the Assumptions 3.2.4, 3.2.5, 3.2.6 and 3.2.7, a service demand d is defined as:

$$d = (x, y, qos, dur)$$

where $x \in \mathcal{N}_I$ is the source node, $y \in \mathcal{N}_J$ the destination node, qos the vector expressing the minimal required bandwidth, β^r, the required end-to-end delay e^r, and the required service level l^r, i.e., $l^r \in Q = \{Gold,\ Silver,\ Bronze,\ Premium\}$, and finally dur the duration of the service. Based on the selected service level, the bandwidth requirement is considered more or less stringent. Indicating with β^o the bandwidth that the provider is expected to offer, it follows that:

- If $l^r = Gold$, then the offered bandwidth β^o has to satisfy the following condition:

$$\beta^o = \beta^r \tag{3.4}$$

  The offered end-to-end delay e^o must fall into a predefined time interval: $e^o \in [d_G^{min}, d_G^{max}]$ with $d_G^{min} < d_G^{max}$ and $e^r \geq d_G^{max}$.

- If $l = Silver$, then β^o has to satisfy the condition:

$$\beta^o = \beta^r(1 - \epsilon'_S) \tag{3.5}$$

  with $0 < \epsilon'_S < 1$ being a positive constant value. The offered end-to-end delay e^o must fall into a predefined time interval: $[d_S^{min}, d_S^{max}]$, with $d_G^{max} \leq d_S^{min} < d_S^{max}$ and $e^r \geq d_S^{max}$.

- If $l = Bronze$, then β^o has to satisfy the condition:

$$\beta^o = \beta^r(1 - \epsilon'_B) \tag{3.6}$$

  where ϵ'_B being a positive constant value with $0 < \epsilon'_B < \epsilon'_S < 1$. The offered end-to-end delay e^o must fall into a predefined time interval: $[d_B^{min}, d_B^{max}]$, with $d_S^{max} \leq d_B^{min} < d_B^{max}$ and $e^r \geq d_B^{max}$.

- If $l = Premium$, there is no specific QoS requirements (i.e., best effort).

A service demand is refereed to as a *multi-provider service* demand whenever the source and the destination nodes belong to distinct providers networks. In real networks, multi-provider service demands may be anything from a video-conference or a video-on-demand application to a virtual link in a Virtual Private Network.

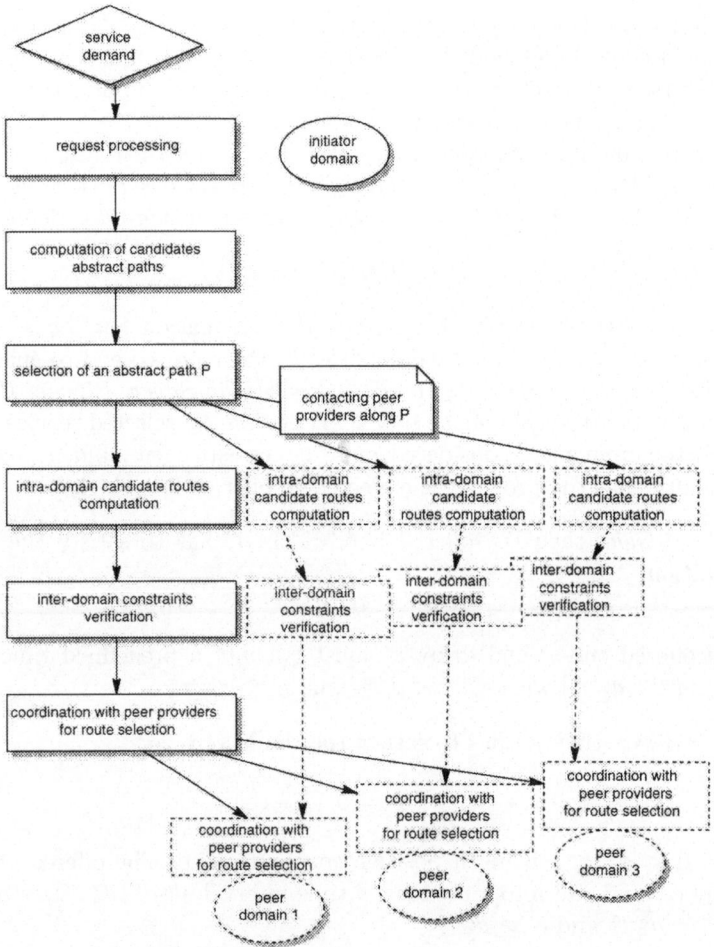

Figure 3.5: *The process of solving multi-provider service set up is decomposed in several basic steps. Every provider involved in the service demand allocation is expected to start intra-domain routines such as computation of candidate routes, including prices, and inter-domain coordination with adjacent networks along the selected abstract path. For the sake of clarity these are omitted in this figure.*

3.2.5   Problem Formulation

The process of allocating a route for a given service demand spanning several provider networks, i.e., the MuSS problem, has been formally expressed in Section 1.1.1. Given the simplifying assumptions presented in this chapter, it is possible to refine the problem this work aims to solve as follows:

> **Given** a multi-provider network scenario consisting of interconnected
> providers networks composed of nodes and bidirectional links, where
> each link is characterised by a given amount of bandwidth and a certain
> end-to-end delay, and an incoming multi-provider service demand, spec-
> ified in terms of source and destination nodes, bandwidth, end-to-end
> delay, duration and service level requirements:
> **Find** on-demand the set of end-to-end routes that satisfy connectivity
> constraints, bandwidth and end-to-end delay requirements according to
> the specified service level over all distinct providers networks involved.
> **Select** one global end-to-end route consisting of specific intra-domain
> routes and inter-domain links in coordination with all the other providers
> involved in the service demand allocation,

Solving the MuSS problem requires therefore addressing a set of issues at
both the inter- and the intra-domain levels. A first important point to remark is
that the network state can change quickly and a certain interval of time may elapse
between the service demand call and its concrete allocation. Therefore, for effectively satisfying QoS requirements, providers need to reserve network resources.
Whenever source and destination nodes of a specific service demand belong to networks controlled by distinct providers, it is necessary to detect how to reach the
final destination network, i.e., which end-to-end path/s will be explored. Hence,
every operator needs to maintain a global view of the multi-provider scenario by
periodically exchanging information with peer providers. For reasons of both scalability and security, providers do not reveal the details of their internal structure,
but they disclose only a summary, or aggregated view, of the costs and availabilities of traversing their domains [91]. The need for aggregation is fundamental
since computation and communication of routing protocols grows at least linearly
in the number of links in the network representation [11]. Moreover, for every
specific service demand to be allocated, a provider has to find an intra-domain
route that can support connection QoS requirements. Along the selected route/s,
resources (i.e., bandwidth) should be optimally allocated to support the required
QoS at a minimal cost. This requires an efficient and flexible way of managing
resources, i.e., of collecting network state information and keeping it up to date,
and a way of finding a feasible path for an incoming demand based on the collected data. In addition, every provider has to specify at which price resources
can be allocated. Pricing a network service is similar to pricing a tangible product, except that the marginal cost of producing the product is replaced by the
opportunity cost of providing the service, which includes both the cost of reserving
and using network capacity. Ideally, prices should dynamically reflect the current
network state and take into account the probability of blocking future incoming
demands (i.e., opportunity cost) when deploying a specific amount of network
resources. This is a very complex task especially in a highly dynamic environment where the network infrastructure should support an efficient routing method
which is able to reflect the offered prices immediately. Current routing algorithms

for large scale networks such as PNNI [91] and BGP [185] do not respond to this
dynamicity need. On one hand, PNNI has not been widely deployed because of
its complexity and because of the missed 'ATM revolution' expected for communication networks. On the other hand, BGP, which is used in the Internet as
inter-domain routing algorithm, selects a route based on relatively static network
information such as network topology [259]. The solution proposed in this book for
providing QoS guaranteed services on a scalable network is an innovative architecture which integrates inter-domain coordination, including inter-domain routing
and peer providers negotiation, intra-domain resource allocation and pricing (see
Chapters 4 and 5).

In order to simplify and better understand the solving approach, the MuSS
problem is further decomposed. Whenever a certain network provider I receives
a specific multi-provider service demand d, the main tasks that need to be performed, after having identified the provider network J that the destination node
is belonging to, are:

- Computing and selecting an abstract path $\mathcal{A}$ interconnecting I and J.

- Contacting all the network providers along $\mathcal{A}$ in order to start the allocation
  process.

- Verifying intra-domain resource availability and constraints in order to define
  the set of possible local routes.

- Making the set of local routes consistent with inter-domain constraints such
  as bandwidth availability, boundary points connectivity, routing policies, etc.

- Selecting a specific intra-domain route satisfying both intra- and inter-domain constraints in coordination with peer providers.

This sequence of steps in which the MuSS process has been decomposed is visu-
alised in Figure 3.5. More details about every step are given in Chapter 4.

3.3   MuSS as a Distributed Constraint Satisfaction Problem

Multi-provider service demand allocation can be considered as a DCSP since the
variables (local network resources such as bandwidth, routers, switches, etc.) are
distributed among agents and since constraints exist among them. More precisely,
end-to-end paths are decomposed into fragments (i.e., distinct variables) corresponding to independent decision makers (i.e., different agents representing distinct providers). In DCSP terms, there is one variable per provider whose values
are route fragments through that provider. Constraints between the variables ensure that route fragments connect and specific DCSP methods are deployed in
order to rapidly access what choices can be part of a consistent end-to-end route,

i.e., a global end-to-end route satisfying both inter-domain and intra-domain constraints. The space of possible solutions provides then the basis for subsequent negotiations.

Before expressing the MuSS problem by making use of the distributed constraint satisfaction formalism, some preliminary assumptions about the communication model which the DCSP paradigm relies upon are given.

Assumption 3.3.1. There is at least one agent for every provider network.

This agent is called *Network Provider Agent* (NPA). Given the number and the heterogeneous nature of various tasks to accomplish for solving the MuSS problem, and considering the existence of networks with a high number of components both hardware and software, a certain number of agents might be working within every provider network. Therefore, every NPA can be intended as the leader or representative of an *agency* consisting of several entities working at different levels in the networks and coordinating each other activity (see Section 8.3 for more discussion on this assumption).

Assumption 3.3.2. Agents communicate using messages.

Messages are encoded in a common *agent communication language* (ACL). The selected ACL has a specific pre-defined syntax and a semantics known to all agents in the framework.

Assumption 3.3.3. The delay in delivering messages is finite, though random.

In real distributed systems, recovery mechanisms and interaction protocols allowing the detection of anomalies should be provided [115].

Assumption 3.3.4. In CSP terms, every provider owns and controls exactly one variable.

What a variable represents in the MuSS context is specified in Section 3.3.1. For large networks different variables would be handled by a hierarchy of coordinated agents controlling distinct parts of the provider domain. Even in that case, the NPI paradigm would still be valid. However, Assumption 3.3.4 simplifies the description of the main mechanisms developed in this work.

3.3.1 The MuSS Constraint Graph

A DCSP is usually represented as a *constraint graph*, where variables are vertices and constraints are edges between vertices. It is fundamental to underline that this constraint graph is not the physical communication network. An edge in the constraint graph is not a physical communication link, but a logical relation between agents. Since each agent owns exactly one variable, a vertex in the constraint graph also represents an agent. The following definition holds in this context:

Definition 3.3.5 (MuSS Constraint Graph). Given a provider set $\mathcal{P}$, a specific incoming demand d and the the selected abstract path $\mathcal{A}$ along which the allocation is started, the corresponding constraint graph consisting of the variables that every

NPA along A and including all the constraints between these variables is called the *MuSS constraint graph*.

The Variables and their Domains

In the NPI framework, the *variable* every provider agent handles is a "local path" (actually it expresses the two end-points of the intra-domain route) specified as a couple:

$$v = (i, o)$$

where i indicates a generic network entry point (or input node) and o represents a generic network exit point (or output node). Both input and output nodes are boundary nodes.

The *values* for each "local path" are all the possible combinations of boundary nodes, i.e., input-output nodes, which represent the possible local routes to allocate to the demand. The set of all possible input-output nodes combinations is the *domain D* for each variable. It is important to underline the difference between *path* and *route*. A path indicates the generic connection between two specific nodes (source and destination), while the route is a specific connection consisting of a sequence of links interconnecting the source to the destination node. Note that only *simple paths*, i.e., loop-free, are considered. In Figure 3.6, the path between a_1 and a_3 can correspond to any of the following routes: $r_1 = a_1 - a_2 - a_3$, $r_2 = a_1 - a_4 - a_3$, $r_3 = a_1 - a_2 - a_4 - a_3$, $r_4 = a_1 - a_4 - a_2 - a_3$.

Consider the example depicted in Figure 3.6. The agent NPA A receives a service demand d for establishing a video-conference between a_1 and b_4. Assume that the selected abstract path is $A - B$. NPA A considers the two following sets:

$$I_A = \{\text{set of possible input nodes in A for allocating } d\}$$
$$O_A = \{\text{set of possible output nodes in A for allocating } d\}$$

Given the demand between a_1 and b_4, $I_A = \{a_1\}$ and $O_A = \{a_2, a_3\}$, since the selected abstract path is $A - B$. The variable for agent A is the couple $p_A = (i, o)$, $i \in I_A$ and $o \in O_A$. The *domain* for v_A is given by the set of all the possible routes connecting i to o: $D_A = \{(a_1, a_2), (a_1, a_3)\}$. Agent B owns the variable $v_B = (i, o)$, with $i \in I_B = \{b_2, b_3\}$ and $o \in O_A = \{b_4\}$. Considering only the points which are directly connected with the predecessor or with the successor along the selected abstract path, this allows the reduction of the complexity of the solving algorithm. The fewer points that are included in I_A and O_A, the fewer possible combinations there exist for allocating a given demand (search space reduction). The domain for agent B is: $D_B = \{(b_2, b_4), (b_3, b_4)\}$. Every path in the variable domain can correspond to one or several routes. For instance, given the path $p_B = (b_2, b_4)$ the set of possible internal routes[8] is

$$\{(b_2, b_3, b_4), (b_2, b_3, b_1, b_4), (b_2, b_5, b_1, b_4), (b_2, b_5, b_1, b_3, b_4), (b_2, b_1, b_4), (b_2, b_1, b_3, b_4)\}.$$

---

[8]Remember that only routes without loops, i.e., *simple*, are considered.

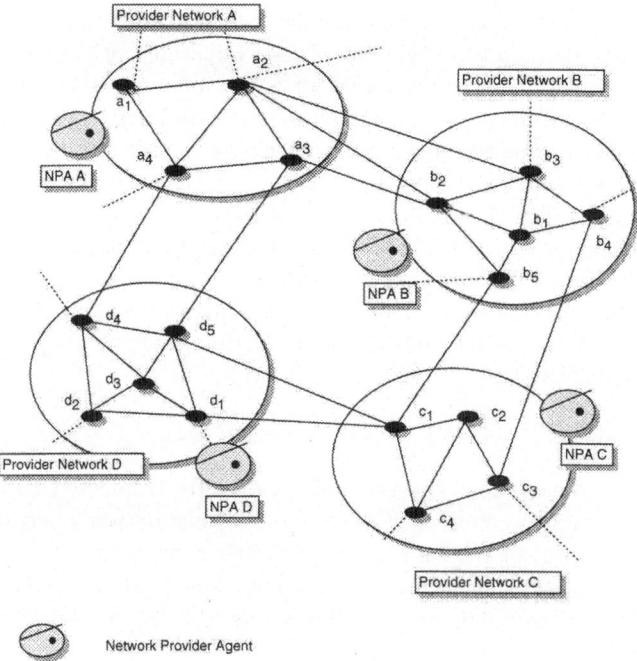

Figure 3.6: *The MuSS process is formalised as a DCSP. The variables (local network resources such as bandwidth, routers, switches, etc.) are distributed among agents and constraints exist among them.*

The MuSS Constraints

In the MuSS context, two main categories of constraints are considered. *Service constraints* correspond to QoS and connectivity requirements as expressed by a specific service demand. *Network constraints* are imposed by network resource availability and providers control and/or management policies [111], [240]. Policies are expressed by inference rules either static or dynamic that every provider applies. Some of these rules depend on which network technology is deployed, others on how available network resources are managed while others on how profit, prices and costs are computed and optimised. Service constraints and network constraints are often inter-dependent. In particular, there are network constraints that can be expressed as a function of service requirements (i.e., the price of a service can be function of the required QoS).

Considering services that span distinct provider networks, QoS and connectivity service requirements first need to be locally translated into constraints between boundary network resources that neighbour providers can use for the inter-domain routing. These *boundary constraints* actually represent the only information that NPAs need to reveal to each other for achieving a globally consistent solution, i.e., an end-to-end route satisfying both inter-domain and intra-domain constraints. Intra-domain topologies, resource availability, prices and internal policies do not necessarily need to be shared for achieving a global consistent solution. These *private* constraints are individually managed and deployed in order to verify and select which internal resources can be used for the service demand allocation. All boundary constraints are binary, i.e., they involve two variables and they refer to the interconnectivity of boundary nodes of adjacent provider networks. Private constraints are unary and must ensure that at least one local path verifies the QoS requirements and the provider policies.

More formally, in order to define the boundary constraints it is necessary to check which are the possible combinations of input/output points from/to each other neighbour network.

$$C(A, B) = \{(o_A, i_B) \mid o_A \in O_A, i_B \in I_B, o_A \rightsquigarrow i_B\}$$

where $o_A \rightsquigarrow i_B$ means that the node o_A is directly connected to the node i_B. $|C(A, B)|$ represents the number of inter-domain links between network A and B. For $|A|$ nodes in network A and $|B|$ in network B there are at most $|A| \times |B|$ possible inter-domain connections, i.e., number of elements in $C(A, B)$.

Considering Figure 3.6, the set $C(A, B)$ of possible combinations of output points of network A and input points of network B is:
$C(A, B) = \{(a_2, b_2), (a_2, b_3), (a_3, b_2)\}$.

All the possible boundary constraints between two providers A and B can be formalised as follows:

$$\mathcal{B}(A, B) = \{((i_A, o_A), (i_B, o_B)) \mid (o_A, i_B) \in S(A, B), (i_A, o_A) \in D_A, (i_B, o_B) \in D_B\}$$

Note that both $C(A, B)$ and $\mathcal{B}(A, B)$ have to be dynamically calculated for every specific demand.

3.3.2 DCSP Formulation

To summarise, in DCSP terms, the MuSS problem can be expressed in the following way:

- The variables are intra-domain paths along the abstract path $\mathcal{A}$ selected for allocating a given demand d.

- The domain D_i of each variable v_i is the set of all intra-domain paths that can be allocated for the demand d inside every network i along the abstract path $\mathcal{A}$.

- There is a constraint on each link forming the intra-domain path (unary intra-domain constraint) ensuring that there is enough available bandwidth on the link.

- There are constraints between two consecutive intra-domain paths along the abstract path $\mathcal{A}$ (binary inter-domain constraints) ensuring that consecutive paths interconnect to each other.

A solution is then a set of intra-domain routes (one for each domain along the abstract path $\mathcal{A}$), respecting intra- and inter-domain constraints. Given a specific service demand, a route is said to be *feasible* when demand QoS requirements and inter-domain connectivity constraints are satisfied. Therefore, solving the MuSS problem means finding the set of feasible end-to-end routes and to select a specific one according to a coordinated interaction of the different providers involved in the allocation.

Given a provider set consisting of distinct interconnected networks, a multi-provider demand $d = (x, y, qos, dur)$, which has to be allocated over a sequence of interconnected local paths represented by the set of variables $\{v_1, v_2, ..., v_j\}$ with $\{D_1, D_2..., D_j\}$ the respective domains containing all the possible paths satisfying both the intra- and the inter-domain constraints:

- Find on-demand the set of feasible routes formed by local interconnected connections so that the intra- and inter-domain constraints on the required network resources are satisfied for each link forming the connection.

- Select one global end-to-end route in coordination with all the other providers involved in the service demand allocation,

The complexity of solving the MuSS problem arises at the intra-domain level because of the need to compute, represent and store the variable domain, i.e., all feasible local routes in a given provider network. Assuming simple and complete networks with n nodes, the total number of routes between any two nodes i and o in the network is equal to:

$$K_{i,o} = \sum_{l=1}^{n-1} \frac{(n-2)!}{(n-l-1)!} \tag{3.7}$$

where l indicates the length of the route. In the worst case scenario, every provider involved in the MuSS process has to compute $K_{i,o}$ local routes and verify which of them are feasible. This can become very expensive especially for large networks (i.e., high n). For this reason, at the intra-domain level, we apply the BI resource abstraction techniques. The BI formalism makes it possible to reduce the complexity of determining local routes satisfying the QoS requirements. BIs provide indeed a powerful clustering technique that can be used to quickly determine the domain of every variable and verify if a route between any two nodes in the network exists before the actual computation of such a route (see next section).

At the inter-domain level, the complexity is mainly due to the need to choose a specific route satisfying both self-interests and inter-domain constraints without revealing strategic information to peer providers. Distributed arc consistency techniques integrated with subsequent negotiations between providers make it possible to simplify inter-domain tasks (see Chapters 4 and 5).

3.4   Network Resource Abstraction

As introduced in Chapter 1, the complexity of solving a specific problem can be reduced by dynamically building abstractions that effectively structure the problem space.

In particular, these kinds of techniques are used extensively in communications networks. Several papers discussed the importance and the effectiveness of resource abstraction for reducing the complexity of routing, control and management tasks (see among others [311], [178], [11]). Additionally, in many deployed networks such as the Internet and ATM networks, routing protocols heavily rely upon the usage of specific information summary, or aggregated view, of the costs and availabilities of traversing network elements and domains. In [46], abstraction methods support reactive strategies for the management of network resources. The aim is to improve the performance of typical network management and control functions by identifying bottlenecks and conflicts. Allemand and Liver [2] propose to build decision support systems in the telecommunication networks area by making use of abstraction techniques. Abstractions between operators and machines (i.e., computers) and between machines and end users are captured and organised as the basis for decision support systems.

In our formalisation of intra-domain tasks such as resource allocation and pricing, the most relevant work concerning abstraction techniques for communication networks is the approach discussed by Frei in [95]. Frei defined a clustering scheme based on blocking islands, which can be used to represent bandwidth availability at different levels of abstraction, as a basis for resource allocation in communication networks. In particular, BI-based techniques can be used to effectively solve off-line QoS routing problems [93]. This approach works by dynamically reformulating an aggregated view of the network state that is the basis of the solving process. How to combine this off-line, centralised techniques with coordination methods to create a distributed use of abstractions capable of solving on-line routing problems inside every communication network domain has been addressed by Willmott [353].

In this context, BI-based techniques are used for creating a dynamic and distributed approach for solving specific issues that concern the MuSS problem. The central idea is to make use of BI methods to quickly assess inside every network the existence of routes between end-points with a given amount of available bandwidth, without having to explicitly search for such routes (see Section 4.3.2). This allows an increase of speed of the coordination process between distinct network providers by pruning out all inconsistent solutions by the intra-domain set of routes considered during negotiation. Moreover, the BI paradigm has been re-elaborated for defining the *Availability Criticalness Evaluation* (ACE) paradigm, which enables the definition of a dynamic *criticalness* factor that indirectly takes into account the probability of blocking future incoming demands by deploying resources on specific links (see Section 5.2.1).

In the following, the main BI definitions and properties that are used within the NPI framework are recalled. For a more extensive and in depth analysis of the BI techniques Frei's dissertation [93] is recommended.

3.4.1 The Blocking Island Paradigm

Blocking Islands (BIs) define a clustering scheme which can be used to represent bandwidth availability at different levels of abstraction, as a basis for distributed problem solving. A β-Blocking Island for a network node x is the set of all nodes of the network that can be reached from x using links with at least β available bandwidth. The Blocking Island definition given by Frei in [93] has been reformulated in this framework so that network links are also explicitly included in every BI. The preliminary definition of *cocycle* is needed.

Definition 3.4.1 (Cocycle). Given a graph $\mathcal{G} = (\mathcal{N}, \mathcal{L})$ and a subset of nodes $S \subseteq \mathcal{N}$, the *cocycle* $\omega(S)$ is the subset of links of $\mathcal{L}$ that have only one and only one endpoint in S.

This definition can be applied to a single network node. Given a graph $\mathcal{G} = (\mathcal{N}, \mathcal{L})$, the cocycle of a single network node $i \in \mathcal{N}$ is the subset of links of $\mathcal{L}$, which i is an endpoint of, and is indicated $\omega(i)$.

Definition 3.4.2 (β-Blocking Island). Given a domain graph $\mathcal{G} = (\mathcal{N}, \mathcal{L})$, a node $x \in \mathcal{N}$ and a bandwidth requirement β, the set $B = (S, I)$, with $S \subseteq \mathcal{N}$ and $I \subseteq \mathcal{L}$, is a β-*Blocking Island* (β-BI) for x if:

1. $x \in S$

2. For all $y \in S \setminus \{x\}$, $\exists$ route $r_i : x \rightsquigarrow y$ with at least β available bandwidth

3. For all $l \in I$ with $l = (a, b)$ (i.e., a and b end-points of l), $a \in S \wedge b \in S$

4. For all $l \in \omega(S)$, $\beta(l) < \beta$ where $\beta(l)$ indicates the available bandwidth on link l.

where $\omega(S)$ is the cocycle of S. Links of $\omega(S)$ are called *outside links* of the β-BI.

Figure 3.7 shows the network graph A and the 64-BI , N_1, for the network node a_1. $N_1 = S \cup I$ consists of the set of network nodes $S = \{a_1, a_2, a_3\}$ and the set of links $I = \{l_1, l_2, l_4\}$. The cocycle of S is $\omega(S) = \{l_3, l_5, l_6, l_7, l_8\}$.

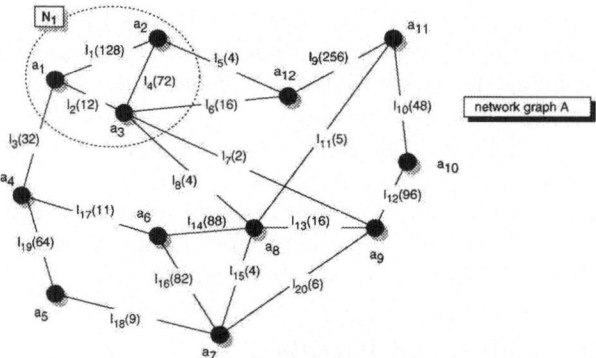

Figure 3.7: *In the network graph A the weights on the links express the available band-width. N_1 is the 64-BI of the network nodes a_1, a_2 and a_3. Link l_2 belongs to N_1 even if it has an amount of available bandwidth that is smaller than 64. The reason is that it is possible to interconnect a_1 and a_3 with $\beta = 64$ by going through l_1, a_2 and l_4.*

It is important to underline that not all the links belonging to a specific β-BI (i.e., the links with both endpoints belonging to the same β-BI) have at least β available bandwidth. If a link l with less than β available bandwidth belong to β-BI it means that there is a alternative route between the endpoints of l with at least β available bandwidth. For instance, in the example depicted in Figure 3.7, the link $l_2 \in N_1$ has an amount of available bandwidth $\beta(l_2) = 12$ that is smaller than $\beta = 64$. However, within N_1, there exists a route interconnecting the end-points of l_2 with at least 64 available bandwidth: $r = a_1 - l_1 - a_2 - l_4 - a_3$.

Given any bandwidth requirement, β-BIs have the following fundamental properties[9].

- BIs partition the network into equivalent classes of nodes.

- BIs are *unique*, and identify *global bottlenecks* or inter-blocking island links. If these are links with low remaining resources, as some links inside blocking islands can be, it means that there is no alternative route to inter-blocking island links with the desired resource requirement.

- The *inclusion property* states that for any $\beta_i < \beta_j$, the β_j-BI for a node is a subset of the β_i-BI for the same node.

- BIs highlight the *existence* and *location* of routes. Frei proves that there is at least one route satisfying the bandwidth requirement of an unallocated demand $d_u = (x, y, \beta_u)$ if and only if its endpoints x and y are in the same

---

[9] All the fundamental properties of β-BIs proven by Frei (see Chapter 3 of [93]) are still valid, given that the only difference between this BI definition and the definition given by Frei is in making explicit the fact that a BI clusters also all the links which both endpoints are in the same BI (see Condition 4 of the Definition 3.4.2).

β_u-BI. Moreover, all the links that could form part of such a route lie inside this blocking island.

3.4.2 Blocking Island Based Network Abstractions

Blocking islands are used to build the β-blocking island graph (β-BIG), a simple graph representing an *abstract* view of the available resources in a domain graph $\mathcal{G} = (\mathcal{N}, \mathcal{L})$: each β-BI is clustered into a single node (abstract node or β-node) and there is an global link (or β-link) between two of these nodes if there is a link in the network joining them. For instance, Figure 3.8 shows the 64-BIG of the provider graph A. An abstract link between two BIs clusters all links that join the two BIs, and the abstract link available resources is equal to the maximum of the available resources of the links it clusters (since a demand can only be allocated over one route). These abstract links denote the critical links, since their available resources are not enough to support a demand requiring β resources. For instance, the abstract link $L_1 = \{l_5, l_6\}$ has at maximum 16 available bandwidth that is the maximum amount of available bandwidth over all the physical links clustered by L_1.

In order to identify bottlenecks for different requirements β_r, a recursive decomposition of BIGs in decreasing order of requirements ($\beta_1 > \beta_2 > .... > \beta_n$) defines a layered structure of BIGs called *Blocking Island Hierarchy* (BIH), see Figure 3.9. The lowest level of the blocking island hierarchy is the β_1-BIG of the network graph. The second layer is then the β_2-BIG of the first level, i.e., β_1-BIG, the third layer the β_3-BIG of the second, and so on. The abstract graph of this top layer is reduced to a single abstract node (the β_n-BI) containing all network resource, since the network graph is supposed connected. The number of levels in the BIH is given by the number of distinct requirements β_r considered for building the BIH. Figure 3.9 shows such a BIH for the bandwidth requirements $\{64, 24\}$. The graphical representation shows that each BIG is an abstraction of the BIG at the level just below (the next biggest resource requirement), and therefore for all lower layers (all larger resource requirements). These abstractions are described by father-child relations. The *father* of a network element (a node or a link) is the β-BIG element (a β-node or a β-link) that abstracts it. The *children* of a β-BIG element are the network elements it abstracts. Notice that the father of a network node is always an abstract node, while the father of a network link is either an abstract link (when the network link is a β-link) or an abstract node (if the network link is clustered in a β-BI).

Definition 3.4.3 (Father Blocking Island). The *father blocking island* of a link $l \in \mathcal{L}$ is the lowest ancestor of l in the BIH of $\mathcal{G} = (\mathcal{N}, \mathcal{L})$ being a blocking island. The bandwidth level of the father blocking island is called the *father level of l*.

Definition 3.4.4 (Father Abstract Link). The *father abstract link* of a link $l \in \mathcal{L}$ is an abstract link L such that: L is either an ancestor of l or l itself, and the father of L is the first blocking island of l.

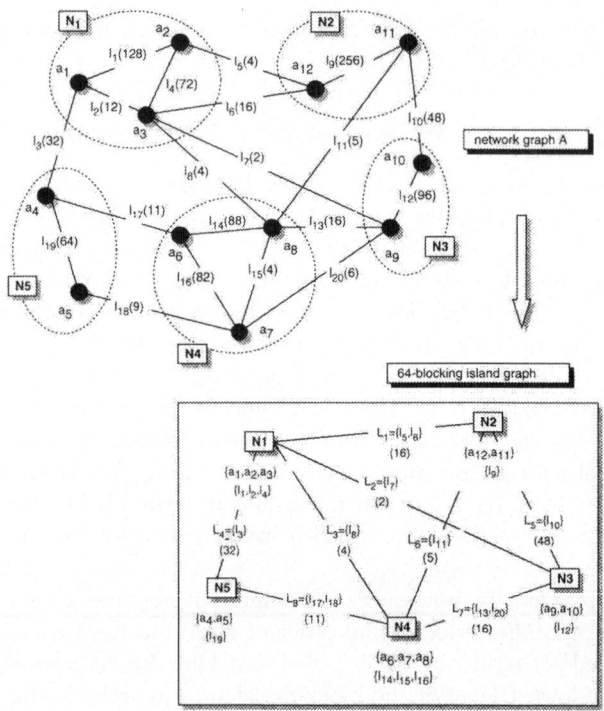

Figure 3.8: *By determining all possible 64-BIs in the network graph, it is possible to build the corresponding 64-BIG consisting of abstract nodes and abstract links.*

These two concepts are equivalent to the definitions given by Frei of *merging blocking island* and *pre-merging link* (see Chapter 3 of [93]). In this context, these concepts have been renamed, since in their original formulation the focus was on merging BI, while here the focus is on detecting the lowest BI (or father) a link is clustered in.

A BIH can be considered as a layered structure of β-BIGs and also as an *abstraction tree* when considering the father-child relations. In the abstraction tree, the leaves are network elements (nodes and links), the intermediate vertices either abstract nodes or abstract links and the root vertex the 0-BI of the top level in the corresponding BIH. Figure 3.10 is the abstraction tree of the BIH in Figure 3.9.

The Cost of Blocking Island

Frei shows in [93] that the β-BI S for a given node x of a network graph can be obtained by a simple greedy algorithm, with a linear complexity of $O(m)$, where m is the number of links in the network. The construction of a β-BIG is straightforward

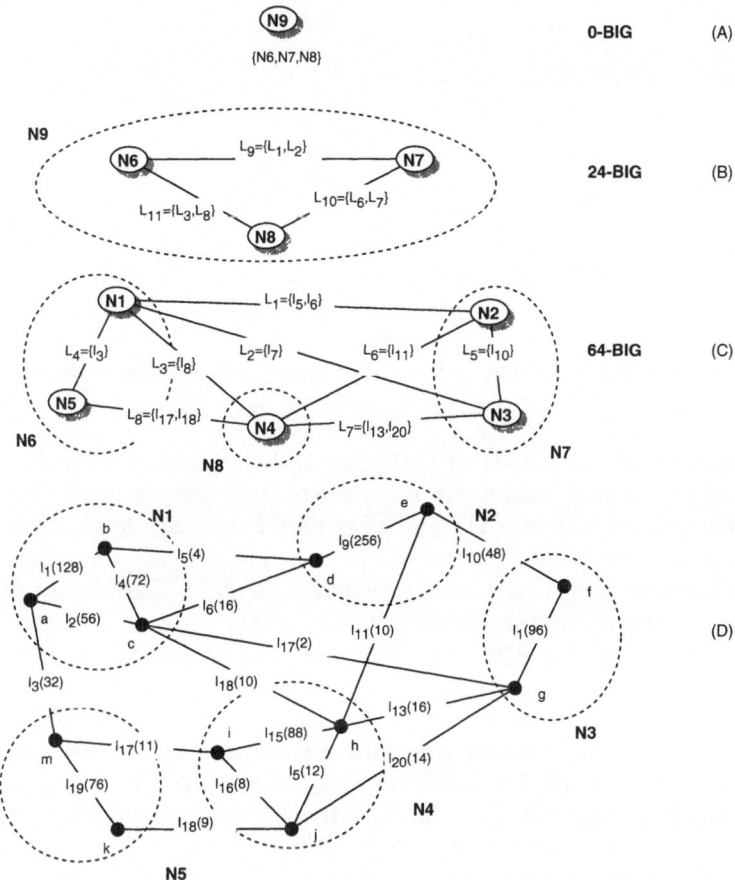

Figure 3.9: *The blocking island hierarchy of network A for bandwidth the requirements {64,24}. The weights (in parentheses) on the links indicate the available bandwidths. (a) shows the 0-BIG. (b) displays the 24-BIG. (c) shows the 64BIG and (d) depicts the network graph.*

from its definition and is also linear in $O(m)$. A BIH for a set of constant resource requirements ordered decreasingly is easily obtained by recursive calls to the BIG computation algorithm. Its complexity is bound by $O(bm)$, where b is the number of different resource requirements. The adaptation of a BIH when demands are allocated or deallocated can be carried out incrementally with complexity $O(bm)$. Therefore, since the number of possible bandwidth requirements b is assumed to be constant, all BI algorithms are linear in the number of links of the network. The set of bandwidth requirements $\mathcal{B}$ is assumed to be discrete and finite, i.e., $b = |\mathcal{B}|$ constant, according to the finite set of possible services that are offered in real networks.

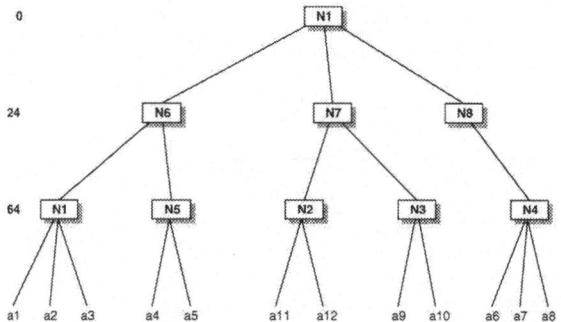

Figure 3.10: *The abstraction tree of the BIH of Figure 3.9 (links are omitted for clarity).*

A BIH contains at most $b(n+1)$ BIs: one BI for each node (n number of nodes in the network) at each bandwidth requirement level plus the 0-BI. In the worst case, there are $\min\{m, n(n-1)/2\}$ links at each bandwidth level, since multiple links between same BIs are clustered into a single abstract link. The memory storage requirement of a BIH is thus bound by $O(bn^2)$. This makes the use of these techniques computationally very cheap.

In order to build a BIH completely independent on the resource requirements of the demands that is only based on the available resources on the links, the *interval* blocking islands hierarchy (Ω-BIH) has been defined [93]. The same properties of the BIH hold, but the main advantage is that there is no need to know in advance the possible bandwidth requirements of the demands in order to build the hierarchy. However, the Ω-BIH requires a higher construction and maintenance overhead (see Chapter 10 of [93]). In particular, the Ω-BIH is more time consuming than the BIH since the complexity is $O(nm)$ instead of $O(bm)$ and usually $n \gg b$. Since one of the major goal of this work is to provide an answer to the MuSS problem in the shortest possible time, the BIH structure (instead of the Ω-BIH) has been selected.

3.4.3 The Differentiated Blocking Island Hierarchy

Given a domain graph $\mathcal{G} = (\mathcal{N}, \mathcal{L})$, the corresponding BIH for the set of offered bandwidth levels $\mathcal{B}^o$ is called Differentiated Blocking Island Hierarchy (Diff-BIH).

The set $\mathcal{B}^o$ is determined in the following way. First of all, given the discrete service model described in Section 3.2.4, a finite set of possible bandwidth requirements is considered:

$$\mathcal{B}^r = \{\beta_i^r\} \tag{3.8}$$

with $\beta_i^r > \beta_{i+1}^r$ and $i \in \{1, \ldots, N\}$, with n a positive constant value. For every service demand d, the amount of required bandwidth is $\beta^r \in \mathcal{B}^r$. Given four different service levels , i.e., gold, silver, bronze and premium, every provider is therefore expected to offer the set of bandwidth levels:

$$\mathcal{B}^o = \{\beta^o_{ij}\} \tag{3.9}$$

with $i \in \{1, \ldots, n\}$ and $j \in \{G, S, B\}$, where G indicates the gold level, S indicates the silver level and B the bronze level. The premium level is not considered since it corresponds to best-effort traffic. Every β^r_{iG}, β^r_{iB} and β^r_{iS} is obtained by applying equations 3.4, 3.5 and 3.6 to every $\beta^r \in \mathcal{B}^r$. The number of levels in the Diff-BIH is given by:

$$b^o = |\mathcal{B}^o| = |\mathcal{B}^r|q$$

where q indicates the number of offered service levels minus one (minus one because for the bronze level no specific bandwidth requirements is fixed). The lowest level of the Diff-BIH is the β^r_{1G}-BIG of the network graph, corresponding to the highest bandwidth requirement. On the top of the Diff-BIH the 0-BIG is reduced to a single abstract node (i.e., the network graph is supposed connected).

As anticipated in Section 3.4.1, the BIH structure is a way of decomposing network resource availability based on different possible bandwidth levels. Whenever two nodes are in the same BI for a given level β of the BIH, it means that there is a route interconnecting the two nodes with at least β available resources (route existence property of BIs, Proposition 3.4 in [93]). However, depending on the granularity of the BIH decomposition (i.e., the number of different levels in the BIH), it may not be possible to detect the existence of a route with a certain amount of resources, even if this route does exist. This happens when the required bandwidth $\beta^r \notin \mathcal{B}^r$. As a main consequence, a demand might be rejected even if there are enough resources available in the network to allocate it. Consider the example depicted in Figure 3.11 case (A). Given the provider network A, its BIH decomposition for the set of requirements $\mathcal{B}^r_x = \{30, 20, 10\}$ and a demand d between a_1 and a_4 with a bandwidth requirement $\beta^r = 6$, it is not possible to detect the existence of a route with enough resources to allocate d by considering its BIH decomposition (obtained for the set $\mathcal{B}^r_x$). The demand would be rejected since nodes a_1 and a_4 are not clustered in any existing β-BI with $\beta > \beta^r$. The same demand d could be successfully allocated building a finer BIH decomposition of network A. Figure 3.11 case (B) shows that with, for instance, an additional level in the BIH, i.e., $\mathcal{B}^r_z = \{30, 20, 10, 8\}$, it would be possible to detect the existence of a route between a_1 and a_4 with enough resources to allocate d, since both a_1 and a_4 belong to the same β-BI with $\beta > \beta^r$. Therefore, a finer decomposition provides more detailed information at the cost of maintaining a structure that is more articulated, i.e., it has more levels. The ideal BIH decomposition should provide all information for demand allocation with the minimal number of levels, i.e., lowest possible complexity.

Definition 3.4.5 (Finest BIH). Given a network I, the *finest* BIH decomposition of I is the BIH that allows the detection of any existing route in the network between any two arbitrary network nodes for any bandwidth requirement β^r.

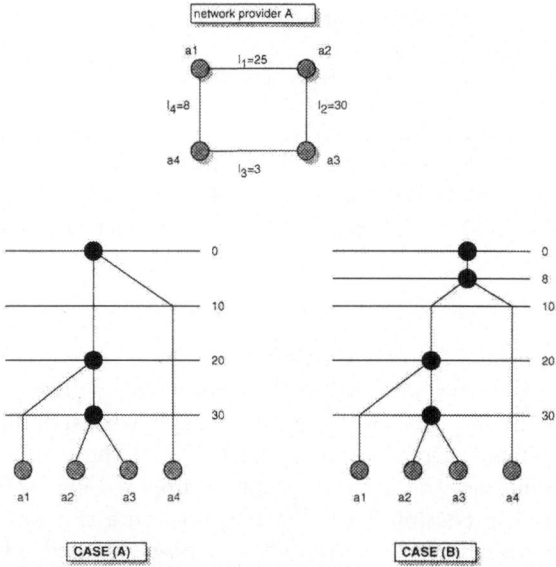

Figure 3.11: *For the given provider network A, two possible BIH decompositions are built. Case (A) shows the BI decomposition built for $\mathcal{B}_x^r = \{30, 20, 10\}$ and case (B) shows the BI decomposition for $\mathcal{B}_z^r = \{30, 20, 10, 8\}$.*

The finest BIH of a given network I is also said to provide full information decomposition of network I.

Proposition 3.4.6. *For any network I with domain graph $\mathcal{G}_I = (\mathcal{N}_I, \mathcal{L}_I)$, the BIH decomposition built by considering the set $\mathcal{B}$ consisting of all possible values of available bandwidth on any link $l \in \mathcal{L}_I$ provides full information decomposition of network I.*

PROOF: For any node $i \in \mathcal{N}_I$, it is obvious that the only possible bandwidth values for which a route starting from or terminating in i exists are superiorly bounded by $\beta_{max} = \max_{l_k \in L_i}(\beta(l_k))$, where L_i is the set of links with i as an end-point. Let us indicate with $\mathcal{B}^i \subseteq \mathcal{B}$ the set of values expressing the available bandwidth on every links $l_k \in L_i$.

In the BIH decomposition of $\mathcal{G}_I = (\mathcal{N}_I, \mathcal{L}_I)$ obtained for the set $\mathcal{B}$, any network node i is clustered inside every β_j-BI, $\beta_j \in \mathcal{B}^i$ because of the inclusion property of the BI (see Proposition 3.10 in [93]). The highest level β in the BIH for which the network node i is clustered inside a β-BI is given by the lowest possible available bandwidth β over all the values in $\mathcal{B}^i$. Therefore, for any bandwidth requirement β^r it is possible to detect if there exists a route in the network, which can accommodate β^r between i and any other arbitrary network node j, by verifying that both nodes belong to the same β_x-BI $\beta_x \in \mathcal{B}^i$ with $\beta_x \geq \beta^r$.   □

Every time a demand is allocated and/or deallocated the set $\mathcal{B}$ changes, therefore, the corresponding BIH should be re-computed in order to provide full information decomposition. This can become quite expensive in terms of computation costs especially for large networks with frequent allocations and deallocations. In order to guarantee full decomposition and reduce the computational complexity, every provider can decide to limit the bandwidth requirements for the incoming service demands. This is what happens in the NPI context.

Proposition 3.4.7. *Given a network I with domain graph $\mathcal{G}_I = (\mathcal{N}_I, \mathcal{L}_I)$ and the finite and discrete set of possible bandwidth requirements $\mathcal{B}^r$, the Diff-BIH of network I provides full information decomposition.*

PROOF: Since for any demand d the bandwidth requirement is $\beta^r \in \mathcal{B}^r$ and the service level is either gold, silver, bronze or premium, the only possible values for which the existence of a route has to be verified are the bandwidth values $\beta \in \mathcal{B}^o$.

By definition the Diff-BIH of a network I is built by considering the set of offered bandwidth levels $\mathcal{B}^o$. Thus, for any node $i \in \mathcal{N}_I$ it is possible to determine if there exists a route with β available bandwidth to any other network node $j \in \mathcal{N}_I$ if $\beta \in \mathcal{B}^o$. This is obvious by construction of the BIH and because of the route existence property of a BI (Proposition 3.4 in [93]). Therefore, the Diff-BIH provides full information decomposition. □

3.4.4 From Blocking Islands to Network Criticalness

The Diff-BIH shows how critical a link is in the network given the current resources availability: the higher the father blocking island of a link is the more critical it is from the resource availability point of view. In the following, a criticalness parameter for any link $l \in \mathcal{L}_I$ in the network graph $\mathcal{G}_I = (\mathcal{N}_I, \mathcal{L}_I)$ of a generic provider network I is formally defined.

Definition 3.4.8 (Link Criticalness Factor). Given a domain graph $\mathcal{G} = (\mathcal{N}, \mathcal{L})$, its Diff-BIH decomposition and a link $l \in \mathcal{L}$, the criticalness factor $\mu(l)$ of link l is given by $1/\beta_F$, where β_F is the bandwidth requirement identifying the father blocking island of link l.

Since the Diff-BIH is the decomposition generated for the set $\mathcal{B}^o$ of bandwidth levels offered, $\beta_F \in \mathcal{B}^o$. Therefore, the higher position a link l is clustered in a BI in the Diff-BIH (i.e., the higher is in the BIH its father BI) the higher is its criticalness factor and therefore its opportunity cost.

One could argue that this simple estimation of the criticalness of links does not properly evaluate the effective probability of blocking future incoming demands in the network. However, a more precise and effective estimation would require to make specific assumptions about the distribution of the future traffic demands in the network. This is out of the scope of this work[10]. In the NPI context, the

---

[10]Especially for a lack of resources and knowledge.

main objective is to demonstrate that correlating the cost of deploying a link to its position in the Diff-BIH has the potential to improve the network resource utilisation and the providers profits. For this purpose and for the sake of clarity we adopted a simple criticalness factor such the one given in Definition 3.4.8.

Proposition 3.4.9. *Given a domain graph* $\mathcal{G} = (\mathcal{N}, \mathcal{L})$, *its Diff-BIH decomposition and two distinct links* $l_x, l_y \in \mathcal{L}$, *if* $\mu(l_x) > \mu(l_y)$ *then* l_x *is more critical than* l_y.

PROOF: This is obvious by the definition of the link criticalness factor (see Definition 3.4.8).   □

When considering two distinct links l_1 and l_2 with $\beta(l_1) > \beta(l_2)$ it is possible that l_2 is less critical than l_1. This happens when the father BI of l_2 is lower in the Diff-BIH than the father BI of l_1. Considering the example depicted in Figure 3.9, the link l_5 is, for instance, less critical than the link l_{13}, since the father BI of l_5 is $N4 \in$ 64-BIG, while the father BI of l_{13} is $N9 \in$ 0-BIG.

The function associating an opportunity cost to any link $l \in \mathcal{L}$ by making use of the criticalness factor $\mu(l)$ is called in this volume the *Availability Criticalness Evaluation* (ACE) cost function. Sections 5.2.1 describe the discrete cost model that has been adopted to define the ACE function and its properties.

By extending the notion of the criticalness of an individual link, the criticalness of a route is defined as follows.

Definition 3.4.10 (Route Criticalness Factor). Given a network I, its domain graph $\mathcal{G}_I = (\mathcal{N}_I, \mathcal{L}_I)$, the corresponding Diff-BIH decomposition and a route r consisting of a sequence of N links $l_i \in \mathcal{L}_I$, the criticalness factor $\mu(r)$ of the route r is given by the maximum criticalness factor value over all the links forming r:

$$\mu(r) = \max_{1 \leq i \leq N} (\mu(l_i))$$

where $\mu(l_i)$ indicates the criticalness factor of a link $l_i \in \mathcal{L}_I$ and N the number of links forming the route.

Chapter 4

The Network Provider Interworking Paradigm

Alice:= "Would you tell me, please, which way I ought too go from here?"
Cheshire Cat:= "That depends a good deal in where you want to get to."
– LEWIS CARROLL, 1865

The *Network Provider Interworking* (NPI) approach provides a set of procedures to achieve automated connectivity across several domains independently of the underlying network technology. On top of this network-to-network coordination layer, it is possible to build mechanisms to automate and improve the way users and providers interact. While user-to-provider interactions are discussed in Chapter 5, in this chapter the focus is on peer providers coordination.

First, the multi-agent system that has been developed for modelling and solving the MuSS problem is introduced. This introduction consists of two main parts:

- The NPI conceptual model including the definition of the different types of agents, their roles, their objectives and their internal architecture (see Section 4.1).

- An overview of the agents interactions used to enable the coordination of self-interested entities in different phases of the service setup process. The main flows of interaction that are discussed in Section 4.2 define how end users and providers converge to agreements that enable service demands to be allocated.

The second part of this chapter (Sections 4.3 and 4.4) goes on to describe the distributed solving process used in the NPI context called the *Distributed Arc Consistency* algorithm. The different phases in which distinct agents acting on behalf of different providers are involved when facing the MuSS problem are examined and specific solutions are proposed. This includes a description of the specific DCSP techniques applied for rapidly accessing what choices (i.e., local routes) can be part of consistent end-to-end routes spanning multiple networks.

4.1   An Agent-based Framework for Service Provisioning

The conceptual model behind the design of the NPI approach is inspired by the network management and provisioning scenario proposed by FIPA in [88]. As shown in Figure 4.1, every final end user involved in the service provisioning process is represented by an End User Agent (EUA). By interpreting user commands and input, every EUA is responsible for contacting providers so that services can be automatically set up. Depending on whether service providers are considered as separate entities from network providers or not, the vertical supply chain can involve several distinct business components: service provider agents (SPAs) and/or network provider agents (NPAs). In this volume, the service providers layer is bypassed since network operators are considered to also act as service providers (which is often the case in real networks). We Adding the SLAs layer as an additional business dimension would require a deep investigation of additional constraints and interests that mainly for a lack of time and resources we decided not to cover in this volume. However, the introduction of more sophisticated brokering and matchmaking capabilities at the service provider level, by considering SPAs as distinct intermediate entities between EUAs and NPAs, could represent a future extension of the NPI paradigm (see Section 8.2.7). Finally, at the bottom of the vertical integration chain, Mask Agents (MAs) inside every network act as gateways between network management and control systems (typically non-agent based) and the agent dimension. The main task of every MA is to convert and translate between low-level technical management primitives and agent communication language (ACL) based messages. For different kinds of underlying network technologies, ad hoc wrappers need to be implemented. It is out of the scope of this volume to address this issue, however, a more detailed discussion is given in Section 8.5. In this work, the data describing the network state in the simulated scenarios is stored in a structure, corresponding to the Management Information Base [128] (MIB), which is dynamically updated during the simulation and which every NPA can directly access.

The MuSS process is started whenever a final end user specifies a multi-provider service demand and communicates it to its personal EUA that can therefore contact a specific NPA. Distinct NPAs have then to coordinate their actions,

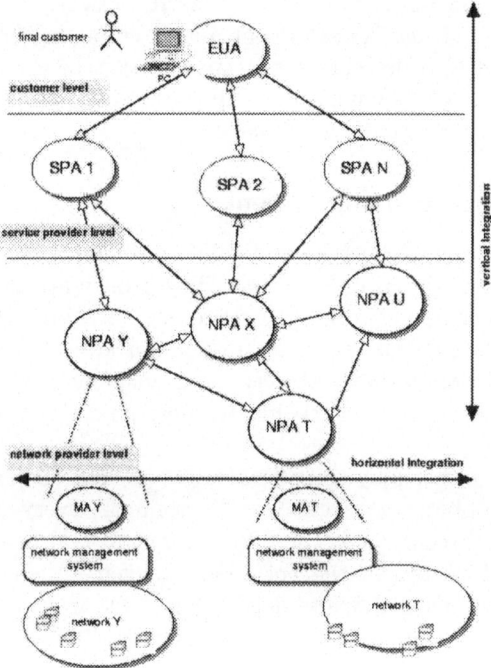

Figure 4.1: *The service provisioning conceptual model: agents interacting on behalf of end users and providers can interact between them in order to automate the service provisioning process.*

since every provider network is managed by a specific entity and all knowledge about network topologies and resources availability cannot be gathered into one agent. Besides the natural distribution of network resources, information, control and interests, there are several reasons for adopting a decentralised solving approach[1]:

- Collecting information in a central point implies important communication costs including significant overheads and can become very inefficient or even infeasible for scalability reasons.

- Centralised structures lack robustness, since any kind of failure damaging the centralised control unit compromises the whole system.

- In settings where different parts are self-motivated, collecting all the information into one agent is not necessarily feasible because of security and confidentiality reasons.

---

[1]A more in depth discussion about fully decentralised versus alternative approaches is given in Chapter 6.

In the NPI context, an agent is considered as a software component that has *perceptions* as input and *actions* as output and is embedded in an *environment*. The environment can also be seen as the specific system the agent is living in. The view an agent has of its environment including other agents populating the scenario is also called the *agent's world view*.

4.1.1   The Network Provider Agent

Network Provider Agents (NPAs) act on behalf of different network providers and play an essential role in the MuSS process. They coordinate the activities taking place inside every specific network they represent with inter-domain tasks that involve external entities. While at the intra-domain level a specific NPA mainly interacts with human operators and network management system components, at the inter-domain level, interactions involve peer NPAs and potential customer EUAs.

Every NPA can either play the role of *buyer* or *seller* depending on whether it is requiring or supplying services. The environment every NPA is embedded in in the experimental scenarios is the NPI system. In real networks, every NPA should be integrated on top of the network management and control platform (see Section 8.5). The main and interdependent goals of provider agents can be summarised as follows:

- Maximisation of the provider profit computed as the difference between the incoming revenue and the current costs incurred for allocating and managing demands.

- Optimisation of the network components utilisation: the main idea is to allocate as much service demands as it is possible given the limited amount of network resources.

- Allocation of service demands so that end user requirements are satisfied. End user satisfaction is a fundamental objective for not loosing customers and therefore future incoming profit.

The internal structure of a NPA consists of several sub-modules, as shown in Figure 4.2. The *input* corresponds to either messages coming from other agents or to end user commands. The *central controller* is responsible for the coordination of several parallel activities, processing inputs and interfacing the agent world with the human operators . The *blocking island* module is responsible for retrieving data characterising the current network state and maintaining the corresponding BI-based representation of available network resources (see Section 4.3.2). In our simulated scenario, the data is stored in a file representing the MIB that is dynamically updated during the simulation. In a real network, the data would be retrieved directly from the network management platform through the use of ad hoc MAs. The NPA *communicator module* parses agents messages, maintains communication channels, controls and manages all ongoing agent conversations. The

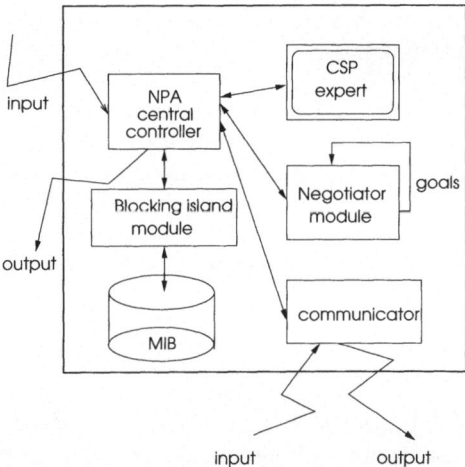

Figure 4.2: *The internal NPA architecture is given by the composition of different modules. Since the activities of these different components are strictly inter-related, a central controller module guarantees their coordination.*

CSP expert unit is a specialised sub-structure responsible for DCSP modelling and for applying typical DCSP consistency and searching techniques. The *negotiator module* generates and evaluates offers that the NPA proposes or receives. More precisely, considering the current state of the network, the various intra-domain constraints, and the selected interaction protocol, this module determines the negotiation strategy. This can require computing prices and profits for specific service demands. Possible *outputs* of the NPA activity are either messages to other agents or actions, such as, for instance, changes in the data configuration, or presentation of options to human operators.

What a Network Provider Agent Needs to Know

The world's view every NPA handles collects information about other agents in the multi-providers network and about inter- and intra-domain network resources. More precisely, this includes:

- A list of all the other network providers in the scenario that a specific NPA can interact with. For every provider in this list, a unique *agent-provider-id* has also to be specified (see Section 3.2).

- A list of all the inter-domain links (see Definition 3.2.9). For every link the set of characteristic parameters, including capacity, delay and estimated cost, are also specified.

- A description of intra-domain resources. This includes all network nodes (internal, boundary and/or shared) and all intra-domain links. Based on this knowledge the NPA computes and maintains the corresponding blocking island hierarchy (BIH) (see Section 3.4.3).

- The set of provider policies. This corresponds to the management, control and routing mechanisms that every provider can apply. In the NPI context, this information represents additional constraints that need to be taken into account when allocating a demand.

How the required information is exchanged between peer-providers in the scenario is presented in Section 4.3.2.

4.1.2 The End User Agent

Within the MuSS process, every EUA plays the role of potential *buyer* of communication services. An EUA receives inputs and commands from the end user is acting on behalf of mainly through a graphical user interface, as depicted in Figure 4.3. This information is then elaborated by the EUA in order to build a profile of the user and formulate specific service demands answering to the end user needs. For efficiently supporting human-agent interactions, two main sub-modules have been implemented [89].

- The *User Dialogue Management Service* component wraps up software components for user interfaces. The main purpose is to translate from/to ACL to/from a human understandable communication language.

- The *User Personalisation* module maintains user models and supports their construction by either accepting explicit information about the user or by learning from observations of the user behaviour.

Depending on the specific type of device the EUA is actually embedded in, the interface between the agent dimension and the human world can be different, e.g., a gesture-based interface to a PDA, a voice interface to a mobile phone, etc. In our simulated environment, the EUA is physically running either on a workstation, therefore, a graphical interface is the most direct and easy way to effectively support the human-agent interaction.

The *central controller* module coordinates the various EUA activities by bridging together all the different agent components. Like for a NPA, the *communicator module* is responsible for parsing agent messages, maintaining communication channels and managing ongoing conversations. The *negotiator module* is responsible for the EUA behaviour during negotiation processes with NPAs. According to the end user profile, the specified service requirements, the utility function, and the selected interaction protocol, the specific negotiation strategy is defined. Possible outcomes of this type of agent are either ACL messages, or actions, such as, for instance, changes in the user profile, questions and/or suggestions to the human end user the agent is acting on behalf of.

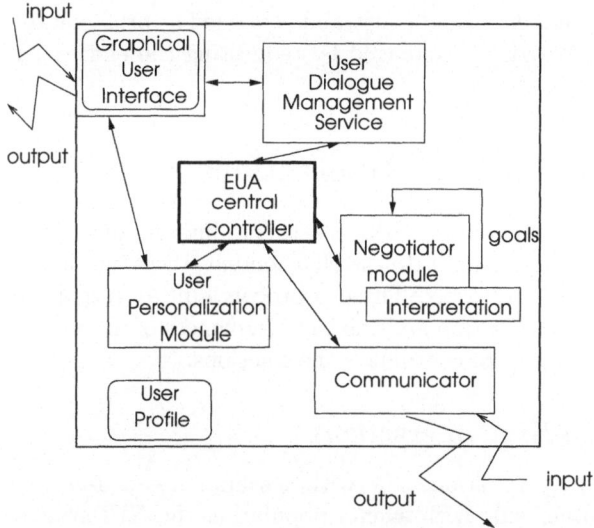

Figure 4.3: *The internal EUA structure consists of several different components. Their coordination is guaranteed by the central controller module.*

The environment an EUA is embedded in is represented by the NPI system. The main interdependent objectives of these kinds of agents are the:

- Concluding agreements with providers for the allocation of service demands. This requires obtaining offers that satisfy QoS requirements and end user constraints and preferences.

- Maximisation of the end user utility computed as the difference between the true evaluation of a certain service and the current price paid to obtain it (see Section 5.1.2).

- Acquisition and maintenance of decision capabilities in relation to the user profile and to the evolution of the environment.

What an End User Agent Needs to Know

An EUA world's view contains information about the specific end user the agent is acting on behalf of and about other agents in the scenario that the EUA might interact with. The main components of this view are:

- A list of all network providers in the scenario that an EUA can contact for the allocation of service demands. This list also includes the agents identifiers needed to specify the recipient/s of the messages sent to other agents.

- The user preferences and constraints encoded as profile of the end user. This profile is created and managed by requesting the end user to enter specific information.

4.2 The NPI Agents Interactions

The MuSS process can be described in terms of interactions taking place between the main entities involved in it. In the NPI context, these interactions are regulated by known standard protocols. These protocols aim at establishing the possible states of a given conversation and the valid sequences of messages that interacting agents can make use of to coordinate their actions.

4.2.1 EUA-to-NPA Interactions

An EUA is interacting on one side with the human end user is acting on behalf of and on the other side with agents populating the NPI system. Human-agent interactions are enabled, as mentioned in Section 4.1.2, by a dedicated graphical user interface. The information an end user enters in this interface is processed by the EUA in order to build the user profile and/or formulate specific service demands that can therefore be submitted to NPAs. Although in the NPI framework interactions between distinct EUAs could potentially take place, for instance, to arrange video-conferences, coordinate plans, schedule meeting, etc., their analysis and dynamics is out of the scope of this volume. In the following, the main focus is on *EUA-to-NPA* interactions enabling multi-provider service setup.

- Every time an end user formulates a request for a service demand[2]$d = (x, y, qos, dur)$, its representative EUA contacts a provider agent in order to get an offer for the demand to be setup. More precisely, the EUA submits a *call-for-proposal* message to allocate d to the NPA that directly controls the source node x. Besides QoS service requirements, the end user can specify specific parameters such as the negotiation timeout and available budget. This information plays a fundamental role in determining the behaviour of an EUA during the negotiation process with a specific NPA.

- A NPA receiving a call for proposal for a multi-provider service demand has to interact with one or several peer NPAs. The mechanism that every NPA is assumed to follow in the NPI context is illustrated in Section 4.3. This mechanism allows the definition of a unique global end-to-end route that is then offered under certain conditions to the EUA.

- Based on the pre-defined negotiation protocol and on the specific negotiation strategy, the EUA examines the proposal coming from the contacted NPA. First of all, the EUA verifies whether the proposal or offer has been received

---

[2]See Section 3.2.4 for the formal definition of a service demand.

within the pre-fixed negotiation timeout or not. Then, an 'on-time' offer is processed for deciding if it is acceptable or not (see Section 4.1.2).

- If the offer is acceptable, the EUA concludes the negotiation by sending an *accept-proposal* message to the corresponding NPA. If the offer is unacceptable, the EUA reaction depends on whether the interaction protocol allows iterations or not. (1) If further no iterations are possible, the EUA can decide whether accepting or not an unacceptable offer by relaxing some service demand requirements. (2) If further iterations are possible, a new call for proposal can be sent to the NPA. Several iterations can occur, if either the EUA makes new call for proposals or if the NPA makes subsequent counter-proposals.

The negotiation protocols applied for regulating EUA-to-NPA interaction processes is described in more details in Section 5.1.1.

4.2.2 NPA-to-NPA Interactions

In the NPI context, every NPA represents a software support (or eventually, in the longer term, a substitute) for human network operators. Therefore, every NPA needs to communicate with (1) human operators, (2) network components (or software management modules controlling these components) and (3) other agents, either EUAs or NPAs. In this section, the focus is on the interactions taking place between peer NPAs. This kind of interaction is triggered by a *call-for-proposal* message to allocate a service demand d that a specific NPA, also called *initiator* (see Definition 3.2.17), receives from a certain EUA in the scenario. Every NPA has an aggregated view of the multi-domain network topology that is needed for computing the abstract paths. An *abstract path* is an ordered list of distinct provider networks between the source and the destination network (see Definition 3.2.16). By choosing a specific abstract path $\mathcal{A}$, a NPA establishes which peer providers will be contacted to start the demand allocation process.

- The *initiator* requests all NPAs along the selected abstract path $\mathcal{A}$ to locally determine the set of possible internal routes for allocating the incoming service demand d. This means that every NPA is requested to check internal resource availability and determine the so called intra-domain or *node consistency* (see Section 4.3.2).

- If all providers are node consistent, i.e., at least one intra-domain route satisfying the QoS requirements of d exists in every network along $\mathcal{A}$, the *initiator* starts the *arc consistency* phase (see Section 4.3.2). All involved NPAs exchange information about inter-domain constraints so that incompatible network boundary resources are discarded from the search. In this way, every NPA reduces the set of possible local routes to allocate d. The main purpose of this phase is to narrow the space of possible choices before actually performing peer-negotiations (i.e., specific local routes selection).

- If the arc consistency interaction terminates successfully, i.e., all NPAs have a non empty set of local routes consistent with inter-domain constraints, the negotiation for selecting a specific end-to-end route is started. The *initiator* runs bi-lateral negotiations with every NPA along $\mathcal{A}$. Every contacted NPA elaborates an offer consisting of a specific local route with certain QoS characteristics at a fixed price. Individual negotiations are more easily integrable within existing network control and signalling mechanisms than, for instance, auctions. In addition, from the initiator perspective this enables for every different interaction the adoption of different policies.

- The *initiator* evaluates all the received offers and elaborates possible global end-to-end offers for the end-user. The NPA-to-EUA negotiation is successful if the global offer is accepted by the EUA (see previous section). The *initiator* confirms to the NPAs the results of the transition. If the negotiation fails and the end-user does not modify its requirements, the demand request is rejected and the initiator notifies all peer NPAs.

NPA-to-NPA interactions are regulated either by the FIPA request, the FIPA query or the *iterated bargain* protocols.

4.3   Coordination of Providers: The Distributed Solving Process

The mechanisms developed for enabling the coordination of distinct providers make use of specific constraint satisfaction techniques and largely rely upon the DCSP-based model of the MuSS problem presented in Section 3.3.

4.3.1   Preliminary Concepts

A binary CSP is defined by the triple (V, D, C) where V is the set of variables, D the set of finite discrete variable values (also called domain) and C the set of binary constraints restricting the values that variables can simultaneously take. The MuSS process can be considered as a *distributed* CSP since the variables (local network resources) are distributed among distinct NPAs and since constraints exist among them. Therefore, the triple (V, D, C) defining a DCSP consists of:

- Distinct variables $v_i \in V$ that belong to different entities.

- $D = \cup D_i$, with D_i being the domain for every $v_i \in V$.

- $C = \cup C_i$, with being C_i the set of constraints for every $v_i \in V$.

As anticipated in Section 3.3.1, a binary DCSP can be represented by a constraint graph, in which each node represents a variable and each arc represents a constraint between variables represented by the end-points of the arc. Since every NPA is assumed to control exactly one variable (Assumption 3.3.4), every node in the constraint graph also represents a specific NPA.

Definition 4.3.1 (Node Consistency). A given binary DCSP is *node consistent* if for every variable $v_i \in V$ every domain $D_i \in D$ has been reduced to the set of all possible values that satisfy the unary constraints $C_i \in C$ on the variable v_i.

Unary constraints restrict the domain of a variable without reference to any other variable. In the NPI context, this means that every NPA j involved in the allocation of a given service demand with specific QoS requirements is node consistent if the domain D_j of possible values that its local variable v_j (or local path) can take is non-empty and contains only values (intra-domain paths) satisfying unary (or intra-domain) constraints. Unary constraints include both service and network constraints (e.g., QoS requirements, routing policies, see also Section 3.3.1).

Definition 4.3.2 (Arc Consistency). A given binary DCSP is *arc consistent* if for any two distinct variables $v_i \in V$ and $v_j \in V$ forming the arc $a(v_i, v_j)$, for every value $x \in D_i$ there exists at least one value $y \in D_j$ such that all binary constraints between v_i and v_j are satisfied. In this case, also the arc $a(v_i, v_j)$ is said to be *consistent*.

It is important to underline that the concept of arc consistency is directional: if an arc $a(v_i, v_j)$ is consistent then it does not mean that the arc $a(v_j, v_i)$ is also consistent. In the NPI system, the *full arc consistency* (or bidirectional arc consistency) is performed: for every couple of variables v_i and v_j from D_i and D_j all values for which either $a(v_i, v_j)$ or $a(v_j, v_i)$ are not consistent are pruned out. This means that if the arc consistency is successful all variable domains are non-empty sets of intra-domain routes satisfying all constraints between neighbour networks (i.e., inter-domain constraints).

If after the arc consistency propagation, at least one variable domain D_i is empty, then the DCSP does not have any solution. If the variable domain size is one for all the variables, then the DCSP has exactly one solution, which is obtained by assigning to each variable the only possible value in its domain. Further search is needed to determine a specific solution, whenever at least one variable domain $D_i \in D$ contains more than one value and all the other domains D_j with $j \neq i$ are non empty[3]. Arc consistency techniques assure the completeness of the NPI solving approach (see Section 4.4.2), since they detect whether a solution exists or not (if a solution exists they can guarantee to find it). If a solution does not exist, it is possible to proceed either by terminating the allocation process or by examining alternative decisions such as relaxing some constraints [370].

Definition 4.3.3 (Feasible Set). Given a binary DCSP the *feasible set* F_i of a variable $v_i \in V$ is the set of all possible values the variable v_i can take, after having after having performed both the node and arc consistency.

---

[3]In the NPI context, the search performed after the arc consistency phase corresponds to NPA-to-NPA negotiations.

In the NPI system, for a given service demand d, an intra-domain path p in a given network j belongs to the feasible set of the variable v_j when both the QoS requirements of d and the network constraints are satisfied. In that case, the path (or route) is said to be *feasible*.

4.3.2   The Distributed Arc Consistency Mechanism

Distinct NPAs need to coordinate themselves in order to find the space of possible end-to-end routes for a given traffic demand, thus creating a concise representation of all possible routes. This space provides the basis for subsequent negotiations about what specific route to select and at which price. In order to converge to global consistent end-to-end routes by revealing only boundary constraints we define the *distributed arc consistency* (DAC) mechanism [35]. The main steps of the DAC process can be summarised as follows:

- **Step 1: Inter-domain source routing**. Computation and selection of a specific abstract path $\mathcal{A}$ along which the allocation is started. *Source* routing means that the forwarding path for all data traffic generated by an incoming demand is computed at the source network, i.e., by the *initiator* NPA. If no feasible abstract paths exist, the EUA is requested to relax the service requirements otherwise the service demand is rejected.

- **Step 2: Peer providers coordination**. Contacting all network providers along the selected path $\mathcal{A}$. The *initiator* requires peer NPAs to participate to the allocation of a multi-provider service demand. If one or more provider/s along $\mathcal{A}$ refuse to collaborate, the *initiator* can decide to investigate an alternative abstract path when possible.

- **Step 3: Intra-domain resource check**. Local resource availability check, also called *node consistency phase*. This step is performed for determining the set of possible intra-domain routes given the current network state. If one NPA along $\mathcal{A}$ does not have enough available resources, an alternative abstract path can be explored.

- **Step 4: Arc consistency filtering phase**. Making the set of intra-domain possible routes consistent with boundary constraints. If one NPA along $\mathcal{A}$ has an empty feasible set, a failure message is sent to each agent involved in the allocation. The *initiator* can decide to investigate an alternative abstract path when possible.

- **Step 5: Negotiation between peer providers** in order to converge to a specific consistent end-to-end route. Among all possible consistent solutions in its feasible set, every self-interested NPA proposes a specific local route without having to reveal internal preferences, strategies or utility functions to any other entity. Chapter 5 focuses on this phase.

Before a more detailed description of the various steps listed above, a last re-mark concerns the degree of "cooperativeness" of the interacting NPI entities. Provider agents can be considered *cooperative* since they exchange a minimal amount of information to find, within every separate network, the set of feasible local routes. Inter-domain constraints represent the information exchanged and revealed to each other for establishing arc consistency (step 4 of the DAC mech-anism). However, during subsequent negotiations to decide what specific route to select and at which price (step 5) self interests prevail, i.e., every agent chooses the best local feasible route for itself. In this sense, therefore, NPI agents can be considered *self-interested* agents behaving in a cooperative way in order to improve their local performance [179].

Maintaining a Global Network View

A preliminary step for every NPA to be able to run the DAC mechanism is main-taining a global view of all peer providers in the scenario. This global view cor-responds to an image of the provider set $\mathcal{P}$ (see Definition 3.2.21) every network provider i is included in.

Definition 4.3.4 (Global Network View). For a given provider network o, the amount of information describing the provider set $\mathcal{P}$ that o is part of represents the *global network view* that o has of $\mathcal{P}$. This includes the complete list of all global nodes and global links forming $\mathcal{P}$.

In order to maintain the global network view, all NPAs in the scenario period-ically exchange messages by following a link-state protocol [17, 230]. Every NPA broadcasts to all other NPAs in the provider set $\mathcal{P}$ the aggregated information summarising the global characteristics of its local network state. More formally:

Definition 4.3.5 (Boundary View). The information exchanged by different net-work providers in order to maintain a coherent global network view forms the so called *boundary network view*. This information includes:

- The node traversal cost and delay that characterise the global node repre-senting the network every NPA controls (see Definitions 3.2.22 and 3.2.23).

- The cost, delay and available bandwidth of the global links every network is directly connected to (see Definitions 3.2.25, 3.2.26 and 3.2.27).

In link-state protocols the main idea is that every network provider broad-casts its boundary view to all the other operators in the provider set. This allows every NPA to build a complete global network view that gives the preliminary information needed for the inter-domain routing process.

In addition to the information describing the global network view, every NPA needs also a detailed view of its local network resources.

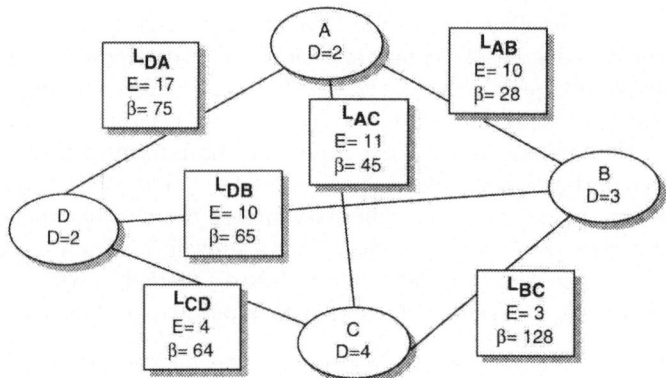

Figure 4.4: *This example displays a provider set consisting of four distinct provider networks where every abstract path is characterised by its delay and its bandwidth and every global node by its traversal delay. Estimated costs are omitted for the sake of clarity.*

Definition 4.3.6 (Local Network View). For a given provider network i represented by the graph $\mathcal{G}_i = (\mathcal{N}_i, \mathcal{L}_i)$, the information describing its network resources, including the delay and the available bandwidth of every link $l \in \mathcal{L}_i$ and the complete list of nodes $n \in \mathcal{N}_i$, forms the *local network view* of i.

In general, the local network view (or state) represents the private information that every provider manages and controls.

Inter-Domain Source Routing

Route computation for a single demand with specific QoS requirements has been extensively considered in the past literature. A good review has been presented by Chen and Nahrstedt in [44]. The computational complexity of this process is primarily determined by the way the metrics used to evaluate the QoS offered by distributed individual network resources (such as links) are computed and combined over a single end-to-end path.

Considering a generic end-to-end path p as a sequence of distinct links l_i, there are two major metrics composition rules that are relevant to the NPI approach:

Definition 4.3.7 (Concave Metric). Let $m(l_i)$ be a metric for link l_i. For any path p, the metric m is said to be *concave* if:

$$m(p) = \min_{l_i \in p} m(l_i).$$

Abstract Path	Hops	Bandwidth	Delay
L_1	A, B	28	15
L_2	A, C, B	45	17
L_3	A, D, B	65	34
L_4	A, D, C, B	64	35
L_5	A, C, D, B	45	36

Table 4.1: *This table lists the possible abstract paths between the two provider networks A and B of the scenario depicted in Figure 4.4.*

Definition 4.3.8 (Additive Metric). Let $m(l_i)$ be a metric for link l_i. For any path p, the metric m is said to be *additive* if:

$$m(p) = \sum_{l_i \in p} m(l_i).$$

When considering a single metric, either concave (e.g., bandwidth, buffer capacity) or additive (e.g., delay, cost) the routing problem can be solved by using either the Dijkstra's [67] or the Bellman-Ford's [16] algorithms with polynomial time complexity. Although, whenever multiple QoS constraints are considered, routing a single demand becomes in general intractable, there are some classes of multiple QoS constraints routing problems that can still be solved in polynomial time [342]. Among these, the central problem of this volume is a *bandwidth-constrained delay-optimisation* routing process. This corresponds to finding a path whose bottleneck bandwidth is above or equal a given value and whose delay is the minimal possible delay.

In the NPI context, an *initiator* NPA computes on-demand, i.e., when the service demand arises, a list of abstract paths satisfying bandwidth requirements and minimising the delay. In general, the optimisation of two metrics simultaneously is an undefined problem, i.e., there is no route that is the optimum for both metrics. Therefore, for a specific demand d, with a given bandwidth requirement β^r, among the routes whose bottleneck bandwidth is greater than β^r, every NPA selects the Pareto optimal [333] route in terms of delay (i.e., lowest end-to-end delay). In this context, a route is Pareto optimal if improving the value for one metric cannot be achieved without making worse the value for at least one of the other metrics.

As specified in Section 3.2.3, the end-to-end delay of an abstract path is the sum of the delays of the global links and global nodes which compose the path itself (see Equation 3.1), while the bandwidth of an abstract path is given by the minimum bandwidth value of its component global links (see Equation 3.2). Before describing the algorithm that has been implemented for the abstract paths computation, some preliminary definitions are needed.

Definition 4.3.9 (Bandwidth Compliance). An abstract path $\mathcal{A}$ between two given networks is β^r-*compliant* if its available bandwidth is greater than, or equal to, β^r.

Considering a specific demand d, with bandwidth requirement β^r, a Pareto optimal route in terms of delay that is at the same time β^r-compliant is said to be a *delay dominant* route. More precisely:

Definition 4.3.10 (Delay Dominance). An abstract path $\mathcal{A}_x$ between two given networks is *delay dominant* if there exists no other abstract path $\mathcal{A}_i$ between the same two networks that is strictly better in terms of delay (i.e., that has lower end-to-end delay) and meets the given bandwidth requirements β^r.

Consider the provider set $\mathcal{P}$ formed by the four provider networks A, B, C and D illustrated in Figure 4.4. Every global node N is characterised by its traversal delay $\mathcal{D}(N)$. For every global link L interconnecting every pair of provider networks two attributes are considered: the estimated delay $e(L)$, which is computed as the average delay of the corresponding physical links delays (see Definition 3.2.25), and the available bandwidth $\beta(L)$, computed as the smallest amount of available bandwidth on the corresponding physical links (see Definition 3.2.27). Table 4.1 lists all possible abstract paths between the provider networks A and B, including bandwidth and delay attributes. When considering a bandwidth requirement $\beta^r = 25$, all the listed abstract path are β^r-compliant and L_1 is the delay dominant path (i.e., lowest delay). When changing the bandwidth requirement, for instance, to $\beta^r = 30$, the abstract path L_1 is not β^r-compliant any more, and the delay dominant solution becomes L_2.

The computation of the β^r-compliant abstract paths minimising the delay is based in the NPI context on a modified version of the Dijkstra's algorithm discussed in [26]. The algorithm every NPA uses takes as input:

- A graph $\mathcal{G} = (\mathcal{N}, \mathcal{L})$, where $\mathcal{N}$ is a finite non-empty set of all global nodes in the provider set $\mathcal{P}$, and $\mathcal{L}$ is the list of global links between any two pair of nodes from $\mathcal{N}$.

- A service demand d between a given source network S and a destination network D with a given bandwidth requirement β^r.

- Equations 3.1 and 3.2 to compute the delay $e(\mathcal{A})$ and the available bandwidth $\beta(\mathcal{A})$ of every abstract path $\mathcal{A}$.

The output is the complete set of β^r-compliant abstract paths from the demand source network node to the specified destination node. The completeness of the algorithm is proved in [26].

The original algorithm proposed by Le Boudec and Przygienda when considering two metrics, one concave (namely bandwidth) and one convex (namely delay), has been shown in [26] to have a complexity of the order of $O(N^3)$ where

N is the number of abstract nodes in the provider set $\mathcal{P}$. The complexity of the procedure used in the NPI framework is reduced to $O(N^2)$, since the shortest spanning tree is not computed for any network node in the provider set $\mathcal{P}$, but only for the specified source network node of the incoming service demand.

Contacting Peer NPAs

Once a specific abstract path $\mathcal{A}_x$ has been selected, the *initiator* contacts all the NPAs along $\mathcal{A}_x$ in order to ask them to collaborate in the end-to-end allocation process of demand d. If the allocation process along $\mathcal{A}_x$ fails, the initiator can iterate the DAC algorithm and investigate a different β^r-compliant abstract path $\mathcal{A}_y$.

If the computation of the abstract paths returns an empty set, it means that there are no paths with enough available bandwidth to satisfy the incoming demand d. In that case, a failure message is sent back by the *initiator* NPA to the EUA. The EUA may decide to relax the bandwidth requirement by diminishing β^r. Then, a *call-for-proposal* message with new service requirements might be sent by the EUA to the NPA.

If several delay dominant abstract paths with different values of available bandwidth exist, the final choice of a specific abstract path is based on the following heuristics:

- *Load Balancing:* The path which has, after having allocated the incoming demand d, the largest amount of bandwidth is chosen.

- *Minimal Number of Hops:* If there are still several options available, the abstract path with the lowest number of hops is selected.

- *Random Balancing:* If after having applied the previous two criteria there are still several options available, a specific path is randomly chosen.

Remark that selecting an abstract path does not mean to specify which boundary nodes the service demand will go through in every domain. The effective selection of specific intra-domain connections forming a global end-to-end route (see Definition 3.2.20) is explicitly done by negotiation during the final step of the DAC mechanism (see Chapter 5).

An alternative approach to the abstract path selection adopted in the NPA system is the parallel search of all possible β^r-compliant abstract paths, by conditionally reserving resources in all of them. However, this approach is practically infeasible due to its excessive operation overhead [45]. For this main reason, an approach that is rather 'sequential' over different possible abstract paths has been selected. The assumption is that an a priori classification given by the aforementioned selection criteria discriminate in which order different paths will eventually be explored. If indeed the allocation succeeds over the first selected abstract path, no further paths will be considered. This implicitly puts the initiator provider in a privileged position (see Section 8.2.4 for more discussion).

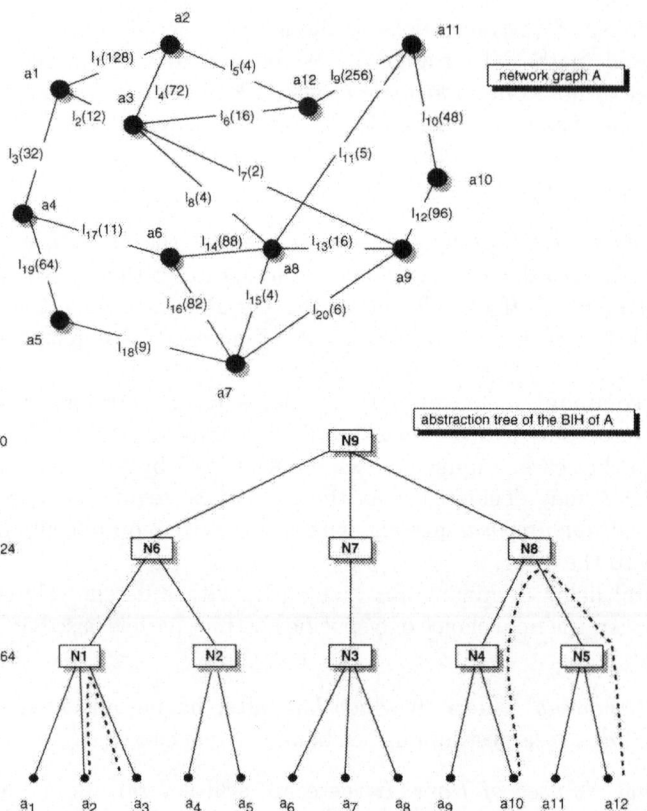

Figure 4.5: *For the network graph A, the abstraction tree of its BIH decomposition allows the visualisation of the existing BIs for different bandwidth requirement levels. In this structure, it is immediate to verify for a given bandwidth requirement whether a route between two specific points exists or not. Whenever two nodes are in the same BI for a given bandwidth level β of the BIH, it means that there is a route interconnecting the two nodes with at least β available resources.*

The Node Consistency Phase

In order to verify if enough local network resources are available to allocate an incoming demand with specific QoS requirements, every provider evaluates its current network state. Given the DCSP-based formalism, this means that every NPA j tests its node consistency (see Definition 4.3.1). This is done by verifying if the domain D_j collecting the possible values that its local variable v_j (or local path) can take is non empty and contains only values (intra-domain paths) satisfying unary (or intra-domain) constraints. Unary constraints include both service and network constraints (e.g., QoS requirements, routing policies, see Section 3.3.1).

In the NPI context, NPAs make use of specific resource abstraction techniques to quickly estimate the current network state. As anticipated in Chapter 3, the used abstraction technique is a clustering scheme based on Blocking Islands (BIs) [93]. The BI approach enables the representation of bandwidth availability at different levels of abstraction in a very compact way. BIs highlight the *existence* and *location* of routes. Frei proves that there is at least one route satisfying the bandwidth requirement of an unallocated demand $d_u = (x, y, \beta_u)$ if and only if its endpoints x and y are in the same β_u-BI. Moreover, all the links that could form part of such a route lie inside this blocking island. This makes it possible to assess the existence of routes between end-points with a given amount of bandwidth, without having to explicitly search for such a route.

The node consistency step is performed by considering the Diff-BIH decomposition of every network (see Section 3.4.2). The Diff-BIH structure is indeed a way of representing network resource availability based on different possible bandwidth levels. Whenever two nodes are in the same BI for a given bandwidth level β of the BIH, it means that there is a route interconnecting the two nodes with at least β available resources (route existence property of BIs, Proposition 3.4 in [93]). Consider the example depicted in Figure 4.5. For the network graph A, the corresponding BIH built for the bandwidth levels $\{64, 24\}$ (remark that the links are omitted for clarity in the picture) consists of nine BIs. Given this structure, it is straightforward to determine that given the bandwidth requirement $\beta = 64$ there is, for instance, a route between nodes a_2 and a_3 since they both belong to the same BI, namely N1. Analogously, it is immediate to detect that there is no route between a_{10} and a_{12} with enough available resources to satisfy the bandwidth requirement $\beta = 64$ since they belong to two distinct BIs, namely N4 and N5. The lowest bandwidth level for which both a_{10} and a_{12} belong to the same BI (i.e., N8) is $\beta = 24$.

In terms of node consistency, given a service demand d requiring β^r available bandwidth, every NPA j controlling variable v_j builds the variable domain D_j by considering all the possible couples of network nodes that belong to the same β-BI with $\beta \geq \beta^r$. Considering the example depicted in Figure 4.5, and the variable v_A between a_1 and a_5 with $\beta^r = 20$, the domain D_A for v_A is:

$$D_A = \{(a_1, a_2), (a_1, a_3), (a_1, a_4), (a_1, a_5)\}$$

This is indeed the set of all possible couples of nodes belonging to the same 24-BI, namely N6, with input node a_1 and output node a_5.

The completeness of the node consistency step is guaranteed by the usage of the Diff-BIH built considering the set of offered bandwidth $\mathcal{B}^o$ (see Equation 3.9). For a given network, its Diff-BIH provides indeed full information decomposition (see Proposition 3.4.7).

The Arc Consistency Phase

One of the main challenges for a process in a distributed system is the detection of a state (i.e., global state) in which a stable property of the system holds [43]. In the distributed arc consistency phase, the system is represented by the group of NPAs along a given abstract path, and the property defining the global state that an agent (process) has to detect is the arc consistency.

This phase of the DAC mechanism allows the elimination of all possible solutions that are not consistent with the given service and network constraints. The main idea is to exchange with peer neighbour providers the minimal amount of information necessary to prune out of the feasible sets the inconsistent choices, or local routes, without revealing strategic and internal data. As anticipated in Section 3.3.1, the exchanged information constraints the possible boundary network resources that can be used to allocate the incoming service demand.

The local arc state each NPA transmits to the *initiator* includes information about the arc consistency of the NPA domain and about the messages that have been exchanged with the neighbours during the arc consistency phase.

Definition 4.3.11 (Local Arc State). Given a service demand d for which a specific abstract path $\mathcal{A}$ has been selected, every NPA x controlling the variable v_x with domain D_x and involved in the arc consistency phase is characterised by a *local arc state* that consists of:

1. A boolean flag , called *localAC*, which is set to TRUE when the local variable domain is not empty and to FALSE otherwise.

2. A local counter, called *localcount*, which indicates the total number of messages received and sent from/to neighbour providers during the arc consistency phase.

The types of messages that are considered for computing the *localcount* value are either the received or the submitted requests from/to neighbours along $\mathcal{A}$ to verify the boundary constraints. For every received request the local counter is incremented by one and for every submitted request decremented by one. At the beginning of the arc consistency, the local counter is set equal to '$|\mathcal{A}| - 1$' for the initiator NPA, and to '-1' for any other NPA along $\mathcal{A}$.

For every received local arc state the *initiator* updates the global arc state.

Definition 4.3.12 (Global Arc State). Considering the allocation of a given a service demand d a specific abstract path $\mathcal{A}$, the *global arc state* of the NPAs along $\mathcal{A}$ consists of:

1. A boolean flag, called *globalAC*, which is set to TRUE if all the NPAs along $\mathcal{A}$ are consistent and to FALSE whenever at least one NPA along $\mathcal{A}$ is not consistent.

2. A global counter *globalcount* that indicates the current total number of requests exchanged by the NPAs along $\mathcal{A}$ to verify the boundary constraints.

The global counter is computed as the sum of all local counters characterising the local arc state of each NPA along $\mathcal{A}$. The effective count starts when consistency messages are exchanged.

The detection of a global arc consistent state along an abstract path $\mathcal{A}$ requires the *initiator* enlisting the cooperation of peer NPAs, which have to record their own local state and communicate it to the *initiator*.

Definition 4.3.13 (Global Arc Consistent State). Given a specific abstract path $\mathcal{A}$, a *global arc consistent state* is achieved when all the providers along $\mathcal{A}$ are consistent and no boundary constraints transit between any of the NPAs $\in \mathcal{A}$.

Proposition 4.3.14 (Consistency of the global arc state). *The global arc state of all NPAs involved in the allocation of a given service demand d along the selected abstract path $\mathcal{A}$ is consistent iff the corresponding boolean flag is TRUE and the global counter is zero.*

PROOF: This is obvious by applying Definitions 4.3.13 and 4.3.12.   □

In the DAC mechanism, the detection of the global arc consistent state is part of the procedure used for propagating boundary constraints. In particular, after a successful node consistency phase, the *initiator* requests all peer providers along the selected abstract path $\mathcal{A}_x$ to start the RunArcConsistency() procedure. For a given service demand d, every NPA x considers as input the variable v_x, its domain D_x, the set of existing constraints (both inter- and intra-domain) C_x and the selected abstract path $\mathcal{A}_x$. Then, every NPA x:

- Initialises the local and the global arc states including both the boolean flags and the counters. The local counter of the initiator NPA along $\mathcal{A}_x$ is equal to the number of providers along $\mathcal{A}_x$ minus 1. For any other NPA along $\mathcal{A}_x$ the local counter is set to -1. In this way, the sum of all local counters is zero at the beginning of the arc consistency.

- Determines the set of inter-domain or boundary constraints B_x that need to be satisfied by the neighbour NPAs.

- Sends to every neighbour NPA along $\mathcal{A}_x$ the set of boundary constraints B_x.

  - If the NPA x is the *initiator*, the set B_x is sent only to its successor NPA[4] along $\mathcal{A}_x$.

  - If the NPA x represents the destination network, the set B_x is sent only to the predecessor NPA[5] of x along $\mathcal{A}_x$.

---

[4]See Definition 3.2.19 for the concept of successor.
[5]See Definition 3.2.18 for the concept of predecessor.

    – If the NPA x represents an intermediate transit network between the source and the destination, the set B_x is sent to the predecessor and to the successor NPAs along $\mathcal{A}_x$.

- Processes every received message m.

    – If $m =$ VERIFY, the NPA X starts the following procedure. For every received set of boundary constraints B_y, the NPA x verifies if all the values in its variable domain are compliant with neighbours boundary constraints B_y. If after this check, D_x is not empty, the NPA x sends a message to the initiator confirming that is locally arc consistent. The value of its local counter is also transmitted. Whenever some values need to be discarded, the constraints set B_x of NPA x is updated and the neighbour NPAs for which the update may be relevant are notified. As a main consequence of the elimination of some values in a specific variable domain along $\mathcal{A}_x$, previously consistent variable domains could indeed need to be revised. Therefore, several iterations might occur before the arc consistency terminates.

    – If $m =$ ARC, the receiver is the *initiator* NPA that verifies if a global arc consistent state has been achieve or not. All the NPAs along $\mathcal{A}_x$ are notified whenever the arc consistency phase is terminated. In the case a global arc consistent state has not been achieved (i.e., *globalAC* = FALSE), the *initiator* can investigate an alternative abstract path. Remark that the selection of an alternative abstract path is optional. This means that the *initiator* may decide to continue trying the allocation of d or notify the EUA that a failure has occurred. In this last case, the EUA may relax some demand requirements and renew the *call-for-proposal*.

    – If $m =$ TERMINATE, the NPA x terminates the arc consistency phase. If *globalAC* =TRUE the negotiation phase can start, if *globalAC* =FALSE, the allocation for the same demand d may go on depending on whether an alternative abstract path is explored or not.

Detecting a Global Arc Consistent State   To better understand the concepts and the mechanisms presented above, let us consider the example depicted in Figure 4.6. The allocation of a demand d between the two network nodes a_1 and c_3 is initiated by the network provider A along the abstract path $\mathcal{A} = A - B - C$. At the beginning of the arc consistency algorithm, every distinct variable domain collects for every distinct network provider the node consistent paths (respectively D_A, D_B and D_C). The set of boundary constraints (respectively B_A, B_B and B_C) are then computed. Every NPA also initiates the local counter. For the *initiator*, i.e., NPA A in the example, the local counter is $localcount_A = |\mathcal{A}| - 1 = 3 - 1 = 2$. For NPAs B and C the local counter is set to -1. NPA A also handles the global

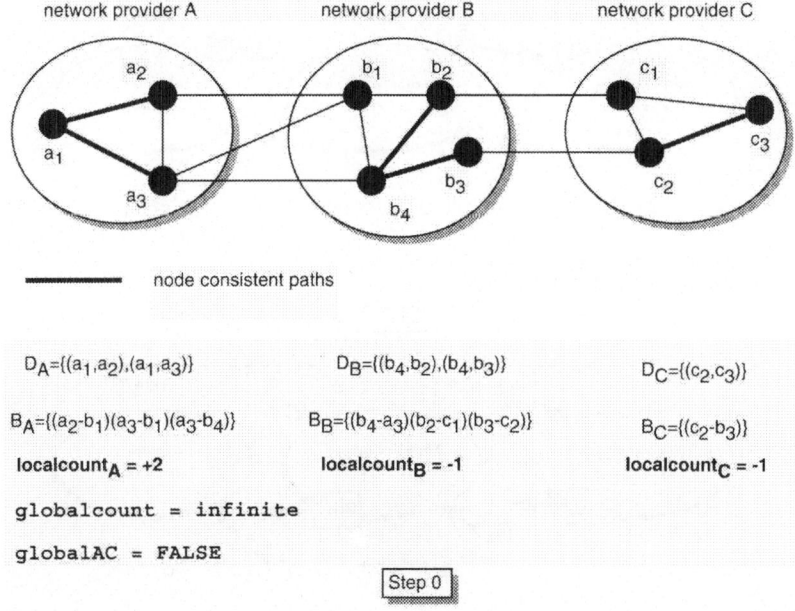

$$D_A=\{(a_1,a_2),(a_1,a_3)\} \qquad D_B=\{(b_4,b_2),(b_4,b_3)\} \qquad D_C=\{(c_2,c_3)\}$$

$$B_A=\{(a_2\text{-}b_1)(a_3\text{-}b_1)(a_3\text{-}b_4)\} \qquad B_B=\{(b_4\text{-}a_3)(b_2\text{-}c_1)(b_3\text{-}c_2)\} \qquad B_C=\{(c_2\text{-}b_3)\}$$

localcount$_A$ = +2   **localcount$_B$ = -1**   **localcount$_C$ = -1**

```
globalcount = infinite

globalAC = FALSE
```

Step 0

Figure 4.6: *The allocation of a demand d between the two network nodes a_1 and c_3 is initiated by the network provider A along the abstract path $\mathcal{A} = A-B-C$. At the beginning of the arc consistency algorithm, every variable domain collects for every distinct network provider the node consistent paths.*

arc state (i.e., global counter and global flag) that is initiated as indicated in Figure 4.6.

The main steps of the RunArcConsistency() procedure identified by the message exchanges taking place between distinct NPAs for detecting a global arc consistent state are illustrated in Figure 4.7. Step 1 shows how every NPA updates its local counter when sending a request to verify its boundary constraints. For every request that is sent to peer providers the local counter is incremented by 1. Upon the reception of requests to verify their local domains with respect to boundary constraints, i.e., Step 2 of Figure 4.7, the NPAs perform several tasks:

- For every VERIFY request message they receive they decrement by 1 their local counter.

- They prune out from their variable domain every value that is not consistent with the received boundary constraints.

- If their variable domain has been updated and their boundary constraints set has changed, they increment the local counter by 1 for every request to verify their new boundary constraints they send to the neighbours.

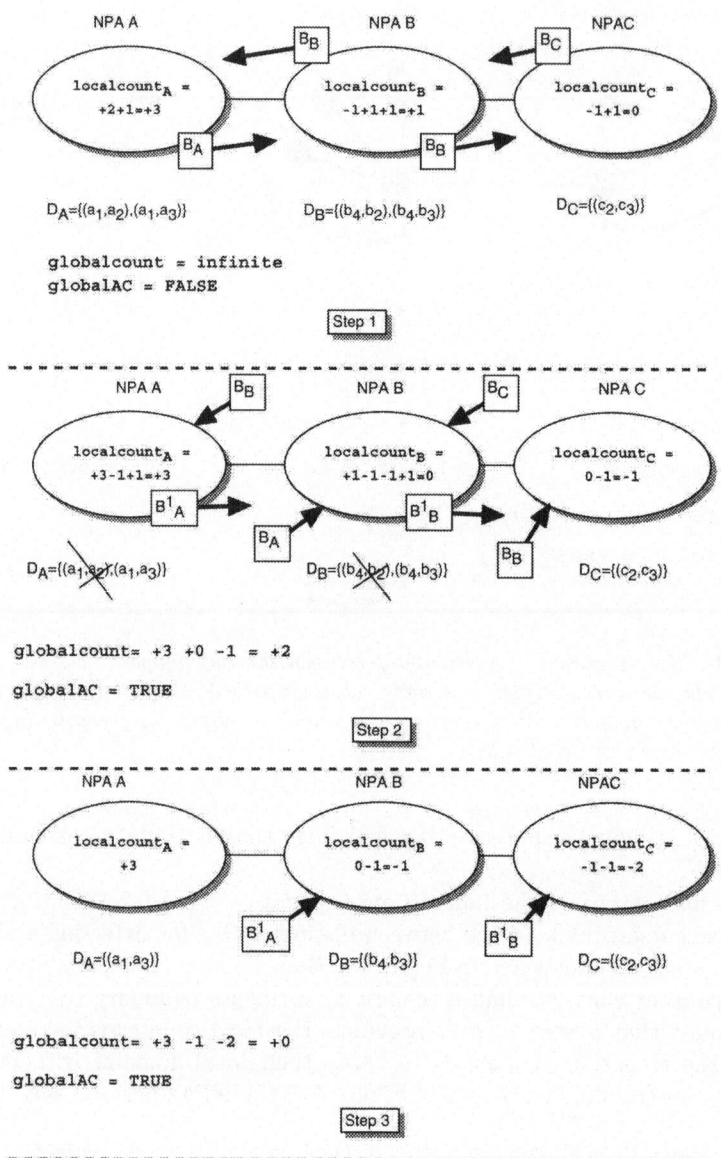

Figure 4.7: *The main steps of the procedure for running the arc consistency are identified by the message exchange taking place between distinct NPAs for detecting a global arc consistent state. Step 1 shows how every NPA updates its local counter when sending a request to verify its boundary constraints. Step 2 shows the different tasks that every NPA performs upon the reception of a request to verify its local domains with regard to boundary constraints. Finally, Step 3 shows that a global arc consistent state is reached since global* $AC = \text{TRUE}$ *and global count* $= 0$.

In this example, the NPA A eliminates the value (a_1, a_2) since it is not consistent with any value in B_B. The NPA B deletes the value (b_4, b_2) since it is not consistent with any value in B_C. The new sets of boundary constraints are respectively $B_A^1 = \{(a_3, b_1), (a_3, b_4)\}$ and $B_B^1 = \{(b_4, a_3), (b_3, c_2)\}$. Remark that the requests to verify the new boundary constraints are sent only to the NPAs for which the change may be relevant. A further task for the *initiator* is the update of the global arc state. The global boolean flag *globalAC* is set to TRUE because both NPAs B and C are locally consistent, i.e., D_B and $D_C \neq 0$. The global counter *globalcount* is obtained by the sum of all local counters. Even if *globalAC* = TRUE the global arc state is not consistent at Step 2, since *globalcount* = +2, i.e., there are still pending requests that have been submitted but not yet processed. Finally, Step 3 shows that since neither further domain updates take place nor further request messages are exchanged between neighbours. The NPAs B and C send their updated local counters to the NPA A that finally notifies that a global arc consistent state is reached since *globalAC* = TRUE and *globalcount* = 0 (see Proposition 4.3.14).

4.4 Efficiency of the DAC Mechanism

Performing the arc consistency for a binary CSP is in general bound by $O(cd^3)$ [192], where c is the number of constraints, and d the upper bound on the number of values in the domain of a variable.

In the NPI context, let $\mathcal{A}$ be the abstract path chosen for the current demand. The number of binary constraints is $|\mathcal{A}| - 1$, that is the number of variables (or involved domains/agents) along the abstract path minus 1. In order to compute the domain size $|D_A|$ of a variable v_A, all the border nodes of the network A must be considered. Let n_A be that number. There are two exclusive cases (remember that source and destination end-points are assumed not to be in the same domain – otherwise no inter-domain routing is required):

1. The network domain contains the source node or the destination node, but not both: $|D_A| \leq n_A$.

2. The network domain is a transit domain, i.e., it does not contain either the source or the destination node, but is part of the abstract path $\mathcal{A}$: $|D_A| \leq \frac{n_A(n_A-1)}{2}$.

Let n be the maximal amount of boundary nodes in the network of a provider along the abstract path $\mathcal{A}$, i.e., $n = \max_{i \in \mathcal{A}} n_i$. Therefore, the complexity of the arc consistency steps of the DAC mechanism is bound by:

$$O(|\mathcal{A}|n^6).$$

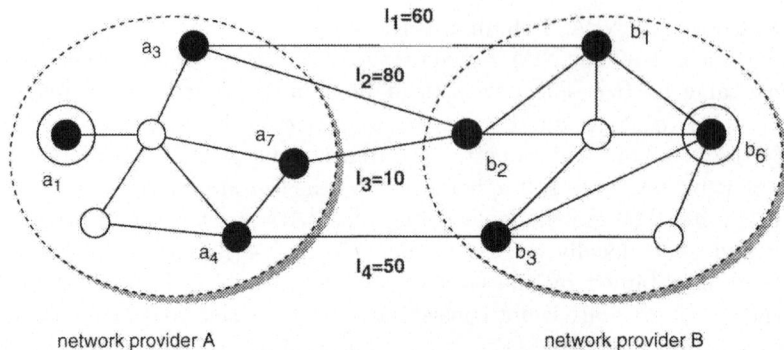

network provider A   network provider B

Figure 4.8: *In this example, two distinct provider networks are involved in the service demand allocation. All network resources, both intra- and inter-domain, are displayed.*

4.4.1   Visualising the Main Steps of DAC Algorithm

The DCSP formalism facilitates the visualisation of the links in the scenario that can or cannot support the required amount of bandwidth β^r. After the arc consistency propagation phase, the only links considered for the demand allocation are the ones which can satisfy both the bandwidth requirements and the inter-domain constraints.

Consider the example depicted in Figure 4.8 and assume that, for instance, network provider A receives a request to allocate a service demand d between the nodes a_1 and b_6 with $\beta^r = 15$. In DCSP terms, NPA A controls the variable v_a that corresponds to a local path of network A. The possible values for v_a are collected into the variable domain $D_A = \{(a_1, a_3), (a_1, a_7), (a_1, a_4)\}$. Analogously, NPA B controls a variable v_b which indicates an intra-domain path of network B. The variable domain for v_b is $D_B = \{(b_1, b_6), (b_2, b_6), (b_3, b_6)\}$. In Figure 4.9 case (A), all the links, both intra- and inter-domain, that satisfy the bandwidth requirement $\beta^r = 15$ are highlighted. This is the configuration after every NPA has performed the node consistency step of the DAC algorithm. During the node consistency phase, NPA B prunes out the local path (b_2, b_6) from the variable domain D_B, since there is no local path connecting the network nodes b_2 and b_6 with enough available bandwidth. This can be verified by checking whether the two network nodes belong to the same β-BI, with $\beta \geq \beta^r$, or not.

Figure 4.9 case (B) depicts the situation after the arc consistency propagation has been performed. At this stage, the variable domains are reduced to the feasible sets of v_a and v_b: $D_A = \{(a_1, a_3), (a_1, a_4)\}$ and $D_B = \{(b_1, b_6), (b_3, b_6)\}$. The value (a_1, a_7) has been pruned out from D_A since it is not consistent with any value in D_B. Remark that the provider B does not reveal the possible values of its variable v_b, but only the boundary constraint which corresponds to not using l_3 as inter-domain link. In this example, there are two possible end-to-end solutions: $(a_1, a_3) - l_1 - (b_1, b_6)$, and $(a_1, a_4) - l_4 - (b_3, b_6)$. The peer provider agents can now

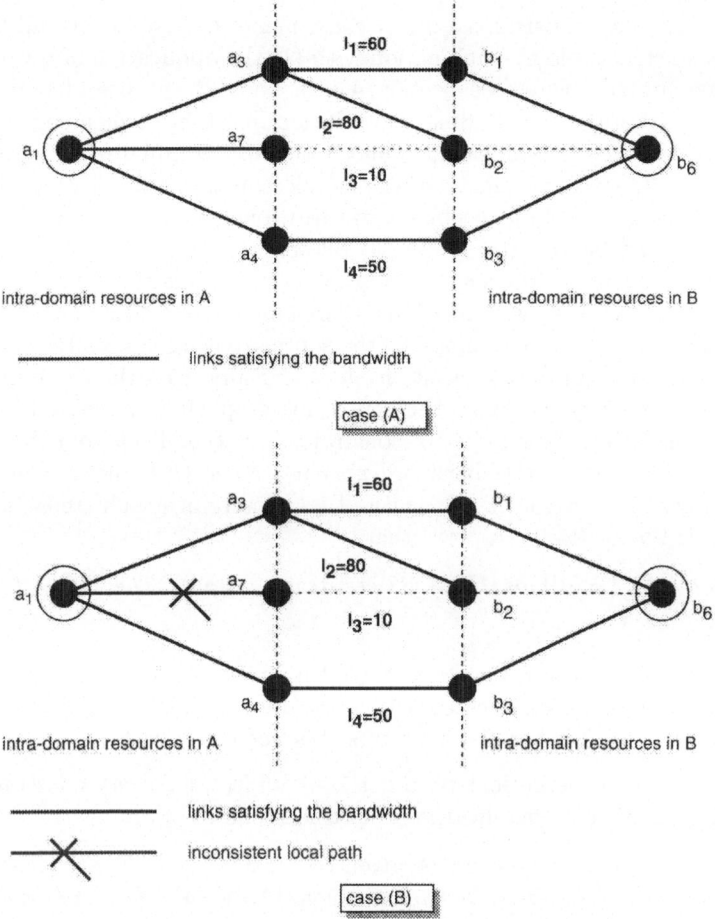

Figure 4.9: *Visualisation of the DAC process: Case (A) shows the links satisfying the bandwidth requirements: this is the situation after the node consistency phase has been performed. Case (B) illustrates the situation after the arc-consistency phase has been performed. The value (a_1, a_7) for variable v_a has been discarded from the variable domain D_A since it is not consistent with any value in the variable domain D_B.*

start to negotiate in order to select which solution to adopt. Section 5.1 focuses on the negotiation phase.

4.4.2   Completeness of the DAC Approach

The arc consistency-based approach is particularly suitable first because it guarantees the completeness of the DAC algorithm , and second because it reveals all the possible solutions without the need to perform any search.

The first characteristics is a direct consequence of how the overall DAC process has been structured in combination with (1) the application of routing mechanisms that are complete (see Section 4.3.2), and (2) the use of node and arc consistency that guarantee finding one solution eventually when solutions exist, and when there exists no solution, to find it out and terminate. The second characteristic is a direct consequence of the simple constraint graph that the DCSP representation of the MuSS problem captures, once an abstract path has been selected. This can be formally proved as follows.

In the NPI context, the constraint graph is an *ordered* constraint graph where the vertices (i.e., NPAs or variables) are ordered linearly from the source to the destination network. This corresponds to the source-routing assumption and means that the order of variable instantiations goes logically from the the source to the destination. For every vertex in the constraint graph it is possible to define its *width* by considering the number of constraint arcs that lead from the vertex to the previous vertices (in the linear order). For a given CSP the maximum of the width of any of its vertices is the width of its constraint graph [169]. In the NPI context, it is thus possible to prove that:

Proposition 4.4.1 (Width of the MuSS constraint graph). *The width of the MuSS constraint graph is 1.*

PROOF: This is obvious by construction of the MuSS constraint graph since once an abstract path has been selected, the constraint graph is a simple chain. Therefore, the maximum number of constraints arcs per vertex is 1.   □

In addition to the notions of 'order' and 'width', for every constraint graph it is possible to define different degrees of consistency.

Definition 4.4.2 (K-consistency). A given constraint graph is *K-consistent* if for any K-1 consistent variables, for any additional Kth variable there exists a value that satisfies all the constraints among these K variables.

Definition 4.4.3 (Strong K-consistency). A constraint graph is *strongly K-consistent* if it is *J*-consistent for all $J \leq K$.

The following theorem proved by Freuder in [97] and [96] is used to demonstrate that the DAC algorithm does not require any search to discover all potential solutions to the MuSS problem.

Theorem 4.4.4 (Freuder). *If a constraint graph is strongly K-consistent, and K strictly greater than the width of the constraint graph, then there exists a search order that is backtrack free.*

Therefore:

Proposition 4.4.5 (Consistency of the MuSS constraint graph). *If the MuSS constraint graph is node- and arc-consistent, then there exists a search order that is backtrack free.*

PROOF: The DAC algorithm uses node and arc consistency to make the MuSS constraint graph strongly 2-consistent. Therefore, by simply applying the theorem 4.4.4 to the strongly 2-consistent MuSS constraint graph, which has width equal to 1 (see Proposition 4.4.1), it is possible to conclude that there exists a search order that is backtrack free. □

Chapter 5

Economic Principles for Agent-Based Negotiations

Ovviamente mi devi una piadina.

– PAOLO PRANDONI, *e-mail 05-09-2001*

In the first part of this chapter, the focus is on the integration of economic principles and mechanisms for enabling agent-based negotiations (see Section 5.1). This includes the definition of the interaction protocols, the decision making models and the possible strategies the NPI agents make use of. Analysis of the provider negotiation behaviour leads to the second part of the chapter in which the main issues of pricing communication services are examined.

As network technologies mature, price tends to become one of the dominant selection factors for purchasers of network services. To compete on prices, providers must reduce the cost of provisioning and maintaining these services. One of the key issues for optimising prices is to efficiently manage network resources and services. In this chapter, after a short background on pricing networks principles and a brief survey of related relevant works (see Section A.1), the *Availability Criticalness Evaluation* (ACE) pricing policy is proposed as an innovative way of evaluating costs and defining prices (see Section 5.2).

While the DAC approach (proposed for the coordination of all agents involved in the allocation of a multi-provider service demand in Chapter 4) explicitly requires the adoption of specific DCSP techniques at both the inter- and the intra-domain level, during the final negotiation phase the requirements on the NPI agents are limited to the use of common negotiation protocols. This means that there are several possible negotiation strategies or pricing policies that each distinct agent involved in the MuSS process may decide to adopt. On the other hand, to guarantee that agents can effectively interact and negotiate, a minimal set of mechanisms need to be fixed and used by all participants in the NPI context (e.g., interaction protocols). What presented in this chapter should therefore be differentiated into two main types of contributions:

- Techniques and mechanisms that are mandatory within the NPI framework for enabling negotiation of self-interested software entities. This includes the specification of the interaction protocols used (see Section 5.1.1) and the multi-provider charging structure (i.e., how providers come up with a global price for a multi-provider service including all expenses in different networks, see Section A.1.2).

- Optional strategies and local decision making methods that either EUAs or NPAs can adopt during negotiation processes (see the second part of Section 5.1). This also includes the way costs and prices of network resources are computed inside each network domain (see Section 5.2).

5.1   Agent-based Negotiations

When trying to solve the MuSS problem, there are two main negotiation flows that can take place:

- The negotiation process between an EUA requesting the allocation of a specific service demand and the contacted *initiator* NPA.

- The various negotiations taking place between the initiator NPA and the peer providers along the selected abstract path $\mathcal{A}$.

The common idea behind both types of interactions is that agents use negotiation for coordination and conflict resolution. This is necessary since distinct agents are autonomous and self-interested. To define the negotiation processes taking place in the NPI framework, some basic principles which have been formalised in the microeconomic theory [166] are used. Quoting Kreps [166], "microeconomic theory concerns the behaviour of individual economic actors and the aggregation of their actions in different institutional frameworks".

The two main types of economic actors considered for the multi-provider service set up process are *consumers* and *firms* of communication services.

Definition 5.1.1 (Consumer). A consumer is an entity that chooses from a set of given offers, selecting the option that maximises its own utility subject to budget constraints and any constraints on feasible consumption.

Remark that the utility is a numerical representation of the consumer preferences. In the NPI context, end users are potential consumers represented by end user agents.

Definition 5.1.2 (Firm). A firm is an entity that is supposed to possess productive capabilities and has as objective function, the profit, which it maximises subject to constraints imposed by its technological capabilities.

The firm profit is usually defined as the difference between the revenue and the costs of producing specific goods or services. In the framework, the main firms are network providers that are represented by network provider agents. The idea is that by using their network resources providers have the capability to 'produce' (i.e., support) communication services.

In the NPI context, interactions between consumers and firms are regulated by automated bi-lateral negotiations. This requires the specification of two fundamental aspects: the *negotiation protocol* and the *negotiation strategy*. The protocol specifies the set of rules that govern the interaction between the negotiation participants, including the possible roles that participants can cover, what deals can be valid and the permissible sequences of actions, i.e., which messages are allowed and in which order. A negotiation strategy establishes the way an agent behaves in an interaction, i.e., what and how specific deals are processed during the negotiation process. Given a certain protocol, there may indeed exist several compatible strategies, each of which may produce a very different outcome. The decision making model that every participant relies upon for achieving its objectives heavily conditions the way the negotiation strategy is followed.

Notice that in the NPI paradigm, both provider and end user agents are assumed to have limited computational resources (i.e., a cost is associated to both computation and communication efforts) and finite time to converge to an agreement (i.e., fixed negotiation timeouts).

Assumption 5.1.3 (Bounded Rationality). For every agent in the NPI scenario there exist a fixed computation cost $c_{comp} \geq 0$ per CPU time unit and a fixed communication cost $c_{comm} \geq 0$ per processed message.

In addition, both end user and provider agents are assumed to be individually rational.

Assumption 5.1.4 (Rational EUAs). An end user agent is assumed to accept a deal only if this does not decrease its own utility.

Assumption 5.1.5 (Rational NPA). A network provider agent is assumed to accept a deal only if this does not decrease its own profit.

5.1.1 The Negotiation Protocol

In the NPI context, bi-lateral negotiations have a limited duration, i.e., a specific negotiation timeout θ_{neg} is fixed. These interactions are regulated by two main different protocols: the FIPA contract net [360] protocol and the *iterated bargain* protocol obtained by modifying the FIPA iterated contract net [361]. In the FIPA contract net protocol, shown in Figure 5.1, one agent takes the role of manager which solicits with *call-for-proposal* messages one or more agents, which

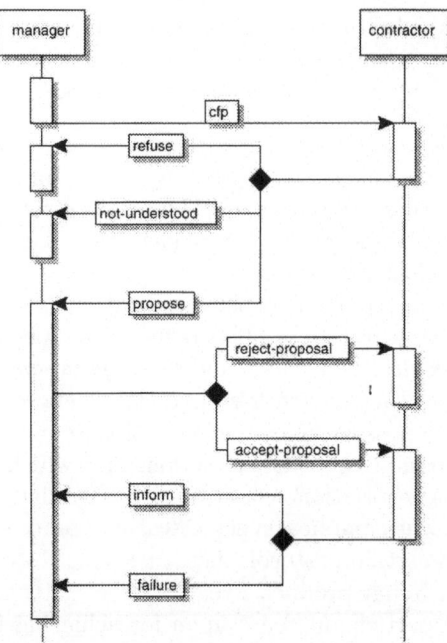

Figure 5.1: *The FIPA contract net protocol.*

assume the role of potential contractors, for performing specific tasks. The potential contractors can then either answer with their bids, in the form of *propose* communicative acts, or decline to make any proposal by sending a *refuse* message. Finally, the manager may accept one or more of the bids, rejecting the others. With the FIPA iterated contract net, the main difference is that the manager has an additional option at this point of the interaction process. It is indeed possible for the manager to iterate the process by issuing a revised *call-for-proposal*.

It has been argued that there is no real negotiation taking place when applying the contract net protocol, since when a conflict arises between the participants it is not possible to bargain [219], [112]. For overcoming this limitation, in the NPI context, a new performative[1] has been introduced in the FIPA agent communication language (FIPA ACL): the *counter-proposal* communicative act. This enables the definition of the *iterated bargain* protocol, which is obtained by modifying the FIPA contract net protocol, see Figure 5.2. From the manager perspective when a proposal does not satisfy the customer requirements, it is possible to either reject the contractor offer and terminate (FIPA contract net), or send a new call for proposal with revised requirements (FIPA iterated contract net), or finally send a counter-proposal offering, for instance, to pay more (iterated bargain protocol). On the other hand, from the contractor side, it is possible to make a counter-proposal

---

[1]The terms *performative*, *message* and *communicative act* are used as synonym in this volume.

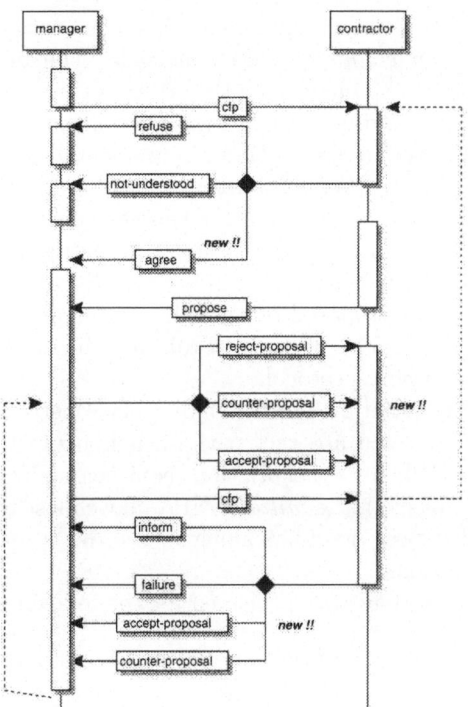

Figure 5.2: *The iterated bargain protocol has been obtained by the FIPA contract net protocol. The main change regards the possibility for both manager and contractor to make counter offers.*

when the potential manager rejects a previously formulated offer. Notice that in comparison with the FIPA protocols, at the beginning of the interaction, an *agree* message that the potential contractor sends to the manager when it accepts to make a proposal, has also been added.

In the negotiation process starting the MuSS process, the EUA formulating a *call-for-proposal* message for the allocation of a specific service demand d is the manager, while the contacted NPA represents a potential contractor. The task to be performed corresponds to providing enough network resources to allocate the demand d so that its QoS requirements can be satisfied. Hence, the initiator NPA starts the DAC mechanism (see Section 4.3.2) in order to be able to generate offers in the form of a *propose* communicative act, which is returned to the manager EUA. The NPA proposal includes the preconditions that the provider is setting out for the demand to be allocated, namely the price and the offered QoS. Alternatively, the *initiator* NPA may refuse to make any proposal.

All offers received by an EUA after the pre-fixed negotiation timeout θ_{neg} has expired are not valid. Among the valid received proposals, the EUA decides if a specific offer is acceptable or not and hence formulates either an *accept-proposal* or a *reject-proposal* or a *counter-proposal* message. The proposals (or counter-proposals) are assumed to be binding on the NPAs, so that once the EUA accepts the proposal, the NPA acquires a commitment to allocate the required network resources for satisfying the demands QoS requirements right away (see Assumption 3.2.6). For this reason, when submitting a proposal (or a counter-proposal), a specific NPA has to reserve the network resources needed to keep this proposal valid. On the other hand, an *accept-proposal* (or a counter-proposal) communicative act is binding on the EUA in that it implies that the end user accepts the price for the required service as proposed by the NPA (or commits to pay the counter-offered amount of money). The protocol regulating the EUA-to-NPA interaction is always the *iterated bargain* protocol.

Concerning the negotiation process taking place between the *initiator* NPA and its peers along a selected abstract path $\mathcal{A}$, it is up to the *initiator* to select which protocol to run. When the negotiation requirements from the EUA are very stringent (e.g., short timeout), the *initiator* NPA may choose the FIPA contract net for interacting with the peer providers along $\mathcal{A}$ in order to reduce the negotiation overhead (i.e., no iterations).

These two protocols have been selected for the following main reasons:

- In real environments, the negotiation between customers and providers is naturally taking place in the form of a bilateral process triggered by a specific request to allocate a given demand sent by a potential customer to a provider[2].

- A simple interaction protocol, such as the *iterated bargain* or the FIPA contract net, can be more easily implemented in real networks since they do not require specific low level signalling techniques. The idea is indeed that agent-based negotiations would be integrated in real networks on top of existing signalling or control mechanisms regulated for instance by protocols such as BGP, RSVP, Control Admission Control (in ATM network).

- FIPA compliant mechanisms increase the potential interoperability of the NPI-based virtual market by proposing standard and common interaction protocols, agent communication languages and ontologies. As mentioned in Section 8.1.1, the *openeness* of the developed coordination paradigm represents one of the main objectives of the work presented in this volume.

Alternative mechanisms such as auctions have also been explored in the communication network context (see Section 2.4.2). However, the main drawback for on-demand service allocation is that providers should reserve in advance too many different portions of network resources. We believe that auctions, for instance, can

---

[2]There are no restrictions for the customer to send the request to allocate the service demand to more than one provider. A one-to-one interaction is considered to simplify the analysis of the mechanisms.

play an important role when *surplus* resources are available to be offered to the market. A concrete example is given by the band-X auction site (www.band-x.com) where network capacity, leased and owned clear channel circuits, dark fiber, wavelengths and ducts are offered.

5.1.2 Modelling the EUA Service Valuation

Every EUA tries to obtain the allocation of a specific service demand at the lowest possible price given QoS requirements and end user preferences. This can be expressed as an optimisation problem where a utility function is used to take into account both service requirements and additional end users conditions. More precisely, the utility is a numerical representation of the consumer preferences that is used to assess the satisfaction of an end user about a specific deal. In the NPI context, it is necessary to preliminary define how end users estimate the value of a certain communication service before formally defining the utility function.

Definition 5.1.6 (Service Valuation). Given a service demand d, the service valuation $\mathcal{V}(d,t)$ is a function of the end user own usage of the service which measures the amount of money the end user is willing to pay at the time t for the allocation of d given specific QoS requirements.

The upper bound for the service valuation is given by the end user *reserve price* $p_r^u(d)$, which represents the absolute maximum budget the end user fixes for the service demand d. The reserve price $p_r^u(d)$ is assumed to be private to the EUA and is used for computing the current service valuation $\mathcal{V}(d,t)$ at any time t as follows:

$$\mathcal{V}(d,t) = p_r^u(d)\sigma(t). \tag{5.1}$$

where $\sigma(t)$ is an increasing function of t with $0 \leq \sigma(t) \leq 1$, $\sigma(0) = 0$ and $\sigma(\theta_{neg}) = 1$, with θ_{neg} is the negotiation timeout. This means that the EUA is likely to pay more as the end of the negotiation process approaches. This kind of behaviour has been modelled by Faratin [82] by using *time dependent tactics*. These tactics model increasing levels of concession as the deadline for the negotiation approaches. In order to specify different risk attitudes of an end user it is possible to vary the convexity degree of the service valuation function. In Figure 5.3, for instance, a linear valuation function is displayed: $\mathcal{V}(d,t) = p_r^u(d)(t/\theta_{neg})$. Section 5.1.4 discusses in more detail the different risk attitudes of end users in the NPI context and the impact this aspect has on the global negotiation outcome.

Another important factor to be considered when defining the service valuation function is the dependency on the QoS requirements. For Premium service demands, i.e., demands that do not require specific QoS guarantees (see Section 3.2.4), the shape of the valuation function, as a function of the bandwidth for instance, is approximated by an increasing and concave function. The concavity is justified by the fact that the increment of the end user benefit is decreasing as the QoS is increasing (see [314], and Appendix of [175]). In the previous literature, the

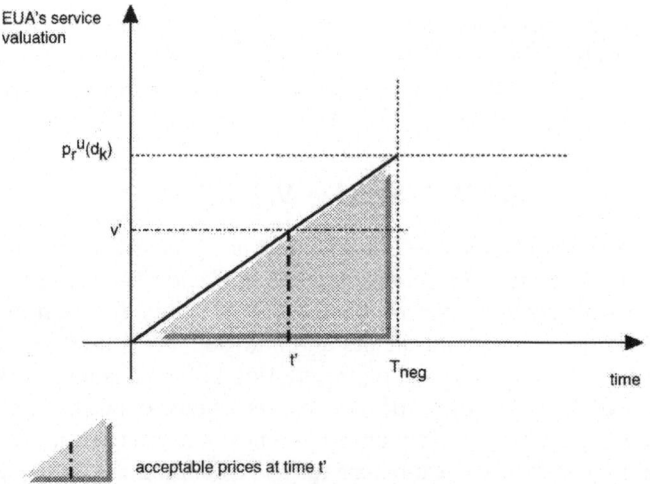

Figure 5.3: *The valuation function is used by the EUA to compute at every instant the specific level of acceptance for a given offer to be accepted. At time t', for instance, the EUA accepts only offers at a price lower than v'. The maximum willingness to pay is bounded by the end user reserve price $p_r^u(d)$ for the specific service demand d. In this example, the valuation function is $\mathcal{V}(d,t') = p_r^u(d)\sigma(t')$ with $\sigma(t') = t'/\theta_{neg}$.*

demand of best-effort services is also indicated as *elastic demand*. For guaranteed services, i.e., Gold, Silver or Bronze service levels in this book, the valuation function is approximated by a step function that has a convex and a concave segment. This is justified by the fact that for these kinds of service there exists a bandwidth threshold below which the end user has a very limited satisfaction (convex segment) and above which the satisfaction rapidly increases up to an asymptotic value (concave segment).

Finally, in the NPI context, the end user utility is formally defined as follows:

Definition 5.1.7 (End User Utility). For a given service demand d, the end user utility $\mathcal{U}(d,t)$ is computed as the difference between the end user valuation $\mathcal{V}(d,t)$ at time t of the service and the price $p(d,t)$ asked by the provider to allocate d:

$$\mathcal{U}(d,t) = \mathcal{V}(d,t) - p(d,t).$$

If the negotiation process between the EUA and the NPA terminates without reaching an agreement, the utility is equal to zero. In the following, to refer to the end user utility for a given negotiation process terminated succesfully, we simplify the notation and we use the expression $\mathcal{U}(d)$, omitting the parameter t.

Notice that what in this book is called the utility has been indicated in other works as the "net benefit" to the end users, and what is called here the valuation has been indicated by these other works as the "utility', see for in-

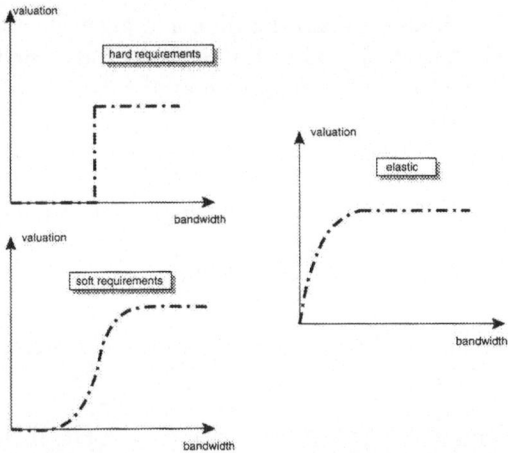

Figure 5.4: *The valuation curves dependency on the offered bandwidth is different for different types of QoS requirements. Demands with hard requirements need specific guarantees for having a value (e.g., traditional telephony). Elastic demands can have some value even if they do not have any QoS guarantee (e.g., Web browsing). The valuation of demands with soft requirements falls in between the two previous ones (e.g., MP3 streaming).*

stance [208], [85], [78]. The terminology used in this volume is coherent with the frameworks discussed in [55], [175], [290] [133].

The EUA Decision Making Process

Designing a negotiation strategy is a task similar to a planning problem[3], since, given the current state of the environment and the agent goals, the main objective is to determine what is the proper behaviour (i.e., what are the most suitable actions to perform) for achieving that goal (or for getting as close as possible to it).

In the NPI context, the decision making model which every EUA relies upon to evaluate offers during the negotiation process with a NPA is based on the constraint satisfaction formalism. Every received offer (or proposal) o for the allocation of a specific demand d, is mapped into a variable v containing three components.

$$v = (p^o, \beta^o, e^o)$$

where p^o is the offered price, β^o the offered bandwidth and e^o the offered end-to-end delay. Each component of v is constrained, from the EUA perspective, by a set of unary constraints.

---

[3]As Hayes and Roth state in [120], planning can be defined as *the pre-determination of a course of action aimed at achieving some goal.*

Definition 5.1.8 (Offer Unary Constraints). For a given demand d, with β^r expressing the required bandwidth and e^r the required end-to-end delay, every offer o received by the EUA at time $t' \leq \theta_{neg}$, is mapped to a variable $v = (p^o, \beta^o, e^o)$ for which the acceptable values are constrained by the set C consisting of the following conditions:

- The offered price must not exceed the current user valuation of the service:

$$p^o \leq \mathcal{V}(d, t')$$

- The offered amount of bandwidth and end-to-end delay must satisfy the following conditions:

    - $\beta^o = \beta^r$ and $e^o \in [d_G^{min}, d_G^{max}]$, with $e^r \geq d_G^{max}$, if the required service level of d is *Gold*.

    - $\beta^o = \beta^r(1 - \epsilon'_S)$ and $e^o \in [d_S^{min}, d_S^{max}]$, with $e^r \geq d_S^{max}$, if the required service level of d is *Silver*.

    - $\beta^o = \beta^r(1 - \epsilon'_B)$ and $e^o \in [d_B^{min}, d_B^{max}]$, with $e^r \geq d_B^{max}$, if the required service level of d is *Bronze*.

    - If the required service level is *Premium* there are no guarantee in terms of QoS.

For every received offer o the *negotiator module*, i.e., a sub-unit of the EUA (see Section 4.1.2), evaluates whether the various unary constraints listed above are satisfied or not. More precisely, the decision making process that determines the EUA behaviour during the negotiation with a provider for the allocation of a given service demand d can be modelled as follows:

- An offer o received at time $t > \theta_{neg}$ is rejected. The utility is equal to zero: $\mathcal{U}(d) = 0$.

- For every offer o received at time $t \leq \theta_{neg}$ the EUA maps o in a local variable $v = (p^o, \beta^o, e^o)$ and verifies if the existing unary constraints on v are satisfied:

    - If all unary constraints C on the values that variable v can take are satisfied the offer o is accepted. The utility is $\mathcal{U}(d) = \mathcal{V}(d, t) - p^o$.

    - When one or more unary constraints C are violated, if it is possible to relax some QoS requirements a *counter-proposal* is sent to the EUA. When no constraint relaxation is possible and the required end user reservation price is violated (i.e., the price is a strong constraint that cannot be relaxed, see Assumption 5.1.4), the demand is rejected. Eventually the EUA can send a new call for proposal after having interacted with the final end user.

5.1.3 Modelling the NPA Decision Making Process

As anticipated in Section 5.1, network operators can be considered as firms that, instead of producing tangible products, provide services. Nowadays, one of the most challenging aspects in all communication networks is how to effectively estimate the marginal cost of supporting these services. There are indeed two main factors influencing the network providers behaviour when allocating a specific service demand d:

- The availability of network resources, so that the required QoS for d can be satisfied, as well as the specific resource control and management policies that apply in the provider domain.

- The cost of allocating resources, which takes into account the opportunity cost of providing this specific service.

Both these aspects influence the price that a provider asks for the service demand d[4]. For this reason, to compete on prices, providers must reduce the costs of network resources allocation and control.

In the NPI context, the total price proposed by the *initiator*[5] to a potential customer EUA, includes expenses for all different providers involved in the service provisioning (i.e., all NPAs along the selected abstract path).

Definition 5.1.9 (Initiator Offered Price). Given a service demand d and a selected abstract path $\mathcal{A}$ along which the *initiator* NPA j is allocating d, the price offered by NPA j for d to the end user is called the *initiator offered price*. This includes expenses for all different providers along $\mathcal{A}$:

$$p^o = \sum_{j=1}^{N} p_j^o(d)$$

where N is the total number of providers along $\mathcal{A}$ and $p_j^o(d)$ the price for the network resources needed to allocate d in every provider network j along $\mathcal{A}$.

The benefit that every provider obtains from the allocation of a service demand d is given by the difference between the revenue and the total cost of the resources needed to allocate d.

Definition 5.1.10 (Provider Profit). Given a service demand d, the provider profit $\pi(d)$ expresses the net benefit for the provider when allocating the service demand d. This is computed as the difference between the revenue $p(d)$ and the total cost $c(d)$ of the resources needed to satisfy d:

$$\pi(d) = p(d) - c(d)$$

---

[4] As MacKie-Mason and Varian pointed out in [190], prices should always reflect costs.
[5] The term *'initiator'* and *'initiator NPA'* are equivalent in this book.

When considering the negotiation process with an EUA, for every allocated service demand d, the revenue $p(d)$ for a provider corresponds to the price that a customer accepts to pay for the allocation of d. The minimal amount of money that the provider is willing to accept for allocating a service demand d is the provider *reserve price* $p_r^p(d)$. Assuming that providers do not sell at a loss, this quantity must satisfy the following condition:

$$p_r^p(d) \geq c(d). \tag{5.2}$$

The reserve price $p_r^p(d)$ is kept private by every NPA for strategic reasons. The cost $c(d)$ comprises the internal resource costs a provider incur, including the communication and computation costs (see Assumption 5.1.3), and the costs eventually incurred for buying resources from peer providers along the selected abstract path for the allocation of d[6].

Therefore, to summarise:

- For any **non-initiator NPA** i, with $i \neq j$, along the selected abstract path $\mathcal{A}$:

  - The cost $c_i(d)$ is given by the sum of communication and computation costs plus the opportunity cost of allocating intra-domain resources for d.

  - The incoming revenue corresponds to the price $p_i^o(d)$ offered by the NPA i to the *initiator* NPA j for allocating intra-domain resources for d.

  - The profit is therefore given by:

$$\pi_i(d) = p_i^o(d) - c_i(d)$$

- For the **initiator NPA** j along the selected abstract path $\mathcal{A}$:

  - The total cost of allocating d is given by:

$$c(d) = c_j(d) + \sum_{i \neq j}^{N} c_i(d)$$

  where $c_j(d)$ is the *initiator* opportunity cost (including communication and computation costs) and N is the total number of providers along $\mathcal{A}$.

  - The incoming revenue is given by the total *initiator* offered price p^o, assuming that this is the amount the end user accepts to pay.

  - The profit is computed as the difference between the incoming revenue and the total cost:

$$\pi(d) = p^o - c(d)$$

---

[6]Section 5.2 gives more details about how prices and costs are estimated in the NPI framework.

The NPA Strategy

When negotiating, a NPA playing either the role of firm or the role of seller aims to maximise its profit. This is done by reducing the costs incurred for providing and managing resources in a way that the offered prices can be reduced. In the current deregulated multi-provider network scenario, price tends to become one of the dominant selection factors for purchasers of network services. Therefore, to keep low prices and maximise the profit the key aspect is to minimise the costs. From a strategic point of view, therefore, when determining proposals for potential customers, every provider faces the following main challenges:

- *Reducing prices* to increase the probability that end users will accept the offer. This makes it possible to satisfy the end users satisfaction and be competitive with respect to other providers in the market.

- *Minimising costs* for maximising the provider benefit while minimising the prices. This means that among the available resources the least expensive should be chosen first.

- *Reducing the risk of future allocation failures.* For this purpose, network resources should be allocated in a way that a good bandwidth connectivity level is maintained.

Hence, for every possible intra-domain route r in the feasible set F[7], which can be selected for allocating a given service demand d, each NPA computes the price p^o (see Section 5.2.1), taking into account both the network state and the service characteristics, and the cost $c(d)$ of using resources along r for d, which includes the criticalness of using these resources (see Section 5.2.1). Notice that among all the computed offers for a given service demand d, the one with the lowest cost c_{min} identifies the provider reserve price $p_r^p(d) = c_{min}$. By associating a price to every feasible route $r \in F$, the NPA defines the set of feasible offers $\mathcal{O}$. In the NPI context, this task is simplified by the used of the BI decomposition of network resources (see Sections 5.2.1 and 5.2.1).

Given the set of possible offers $\mathcal{O}$ to allocate the service demand d, there are two main strategies for a NPA that have been investigated in the NPI context:

- The *Most Profitable First* (MPF) strategy: among all the possible offers in $\mathcal{O}$, the NPA proposes first the offer o that maximises its profit. Then, if more than one offer produces the same profit, the NPA proposes first the least critical route.

- The *Least Critical First* (LCF) strategy: among all the possible offers in $\mathcal{O}$, the NPA proposes first the least critical offer o. Then, if more than one offer have the same criticalness factor, the NPA proposes first the most profitable route.

---

[7]For the concept of *feasible set* see Definition 4.3.3.

If a specific offer o is accepted by the end user, the negotiation terminates and the NPA profit is given by $\pi(d) = p^o - c(d)$, with p^o the price offered by the provider and $c(d)$ the cost of the selected resources. If the offer o is rejected the NPA can make an alternative proposal (that differs depending on the strategy). In this latter case, it may also be possible that the NPA receives (instead of a rejection) a new call-for-proposal with different QoS requirements or a counter-proposal. When the NPA receives a *counter-proposal* asking for the allocation of d at a counter-price (lower than the previously offered price p^o), it verifies if any of the alternative possible offers in $\mathcal{O}$ can satisfy this requirement. If this offer exists, the counter-proposal is accepted and the provider profit is given by the difference between the counter-price and the cost $c'(d)$ of the new selected route. On the other hand, if no other routes have a price lower or equal than the counter-price the counter-proposal is rejected.

The main motivation beyond the MPF strategy is to maximise the provider profit for every allocated demand independently on its criticalness (short term perspective). However, there is no further consideration on whether the most profitable option at the current time does increase or not the probability of rejecting future incoming demands.

With the LCF strategy, the average offered prices are lower. This first of all increases the probability that end users will accept the offers. In addition, because of the usage of the less critical resources, providers are able to allocate resources in a way that a good bandwidth connectivity level is maintained. This reduces the risk of future allocation failures. Therefore, the main motivation behind the LCF strategy can be summarised as follows.

- The lower a BI is in the Diff-BIH, the smaller is the number of links and nodes clustered in it. Therefore, when looking for the least critical route the search space is reduced.

- The lower a BI is, the less critical are the links clustered in it. In this way, a better bandwidth connectivity level is maintained in the network. This has the main effect of reducing the risk of future demand allocation failures.

Experimental results show that providers adopting the LCF strategy are able to allocate a higher number of service demands on average and obtain higher profits (see Chapter 7).

The Complexity of the NPA Strategy

The adoption of a certain negotiation behaviour and a specific strategy have an important impact on the performance of every NPA. This is mainly related to (1) the complexity of the mechanisms developed, which enables offers to be made, and (2) to the amount of information needed to follow a specific strategy.

With the MPF strategy, in order to select the most profitable option, a provider has to compute the profit of every possible intra-domain feasible route

$r \in F$. For a given network with n nodes, the total time complexity is given by $O(fn)$, where $f = |F|$. For every route $r \in F$ there are indeed at maximum $(n-1)$ links, if n is the maximum number of nodes in the network, and for every link the NPA has to compute the price and the cost. Even if potentially the size of F can become quite high when considering large networks[8], the node and arc consistency filtering steps (see Chapter 4) helps in reducing the size of the feasible set F.

The LCF strategy corresponds to the *lowest level* (LL) heuristic discussed by Frei [93]. The idea is to give the priority to the offers for which the corresponding selected route r has the lowest criticalness factor. Given the Diff-BIH decomposition of the network, computing this route has a time complexity of $O(n^2)$ (see Section 5.2.3 of [93]). Besides being cheaper in terms of computational complexity, experimental results confirm that when following the LCF strategy the control of network resources is optimised enabling higher profits for the providers (see Section 7.3.2).

5.1.4 The Negotiation Outcome

A demand is said to be successfully allocated when an agreement between the negotiation participants is found before the expiration of the negotiation timeout. Assuming that every user does not accept to pay a price higher than its own reserve price p_r^u, and that a provider does not offer services at a price lower than its own reserve price p_r^p, an agreement can only be found if $p_r^u \geq p_r^p$.

¿From the customer perspective, the negotiation outcome is evaluated in terms of utility, while, from the provider perspective, the benefit of a negotiation process is estimated by the profit. From a *supra partes* point of view the global benefit for the 'society' is defined in terms of social welfare.

Definition 5.1.11 (Social Welfare). For a given negotiation process involving N distinct providers and M distinct end users, the social welfare is defined as the benefit to the society measured by the sum of the providers profit and end users utility:

$$\mathcal{W} = \sum_{i=1}^{N} \pi_i + \sum_{j=1}^{M} \mathcal{U}_j$$

The strategy every participant adopts during the negotiation process plays a fundamental role not only in influencing the reachable social welfare, but also in determining whether an agreement can be reached or not. On the end user side, the influence on the negotiation outcome is expressed through the EUA risk attitude. The risk attitude of an EUA is directly mapped into the way the service is estimated, which is summarised, in the NPI context, by the valuation function. In particular, given a generic valuation function, $\mathcal{V}(d,t) = p_r^u(d)\sigma(t)$, different users attitudes are specified by different shapes of the curve $\sigma(t)$. For a given

---

[8]Developing the Equation 3.7, it can be found that the upper bound for the size of the feasible set is $f = O(n^{n-3/2} exp^{-n})$.

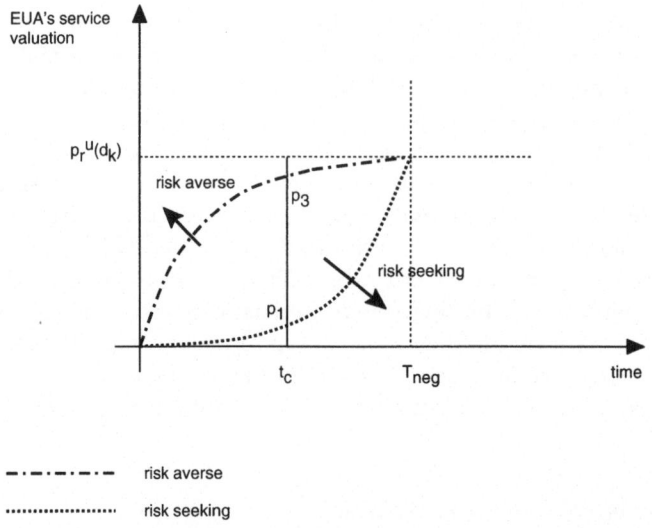

Figure 5.5: *Different shapes of the valuation function incorporate different kind of risk attitude. In this example, different attitudes are represented by different curves.*

demand d, the end user willingness to accept offers at a price closer to the reserve price p_r^u as the negotiation timeout approaches varies depending on the convexity degree of $\sigma(t)$. Therefore, for the allocation of the same service demand d, the same negotiation timeout θ_{neg} and the same reserve price p_r^u, at the same instant t the concession level of agents with different risk attitudes can be very different and determine quite dissimilar outcomes of the overall service set up process. Consider Figure 5.5, depending on the valuation function used by the end user for the allocation of the same service demand d, the same negotiation timeout θ_{neg} and the same reserve price p_r^u, at the same time the concession level can be very different. For instance, at time t_c the concession level p_3 is the highest for the risk averse behaviour, while the lowest concession level p_1 characterises the risk seeking attitude. The concession level of a risk neutral agent is always comprised between the concession levels of more or less risky agents. In all the experimental tests reported in Chapter 7, the risk attitude of the EUAs populating every scenario was randomly determined. The fundamental idea is that a random distribution of the risk attitude of potential customers more closely corresponds to real heterogeneous markets. Additional tests would be necessary to estimate more precisely the relationship between customers risk attitude and sellers pricing policies in different possible settings (see Section 8.3).

On the providers side, the direct influence of a NPA on the global negotiation outcome is determined by the specific strategy in combination with the particular adopted pricing policy. In order to better evaluate the impact of a NPA behaviour on the negotiation outcome, it is therefore necessary to define how prices and costs are estimated in the NPI context. A review of background principles for pricing in communication networks is given in Appendix A.

5.2 The NPI Pricing Model: from Criticalness Costs to Pricing

Pricing a communication service spanning networks owned and controlled by distinct operators involve two main tasks:

- From a global perspective, defining the *inter-domain charging structure*. This means to establish how the global price that an end user has to pay comes out from a combination of all individual prices in distinct networks.

- From a single provider perspective, defining the *intra-domain pricing* approach. This corresponds to the computation of costs and therefore prices inside every network involved in the service demand allocation.

The inter-domain charging structure used in the NPI context, has been inspired by the *edge pricing* model [303] (see also Section A.1.2 in Appendix A). The fundamental idea is that the end user is charged only by the first network provider (i.e., the *initiator*) along the abstract path connecting source and destination nodes. This unique global charge includes expenses for all different providers involved in the service provisioning. The main advantage of this kind of architecture is that the local pricing process can be performed by every provider independently on how others decide to estimate prices. On the other hand, the main drawback from the providers perspective is that they have to come to mutual peer agreements and therefore coordinate each other despite possible conflicting interests. From the users perspective, this approach simplifies the billing and accounting procedure. There is indeed only a bilateral negotiation process and a unique bilateral contract between the end user and the directly contacted provider. In the original edge pricing model, the *initiator* computes the total price for a service by considering static pre-existing bilateral agreements with peer providers. This means that the price is based on the *expected* congestion along the *expected* path and there are no "per-flow settlement payments". In the NPI context, this process has been modified. On-demand peer agreements allow the computation of prices that dynamically reflect the current criticalness of the network resources used along the selected abstract path. Although, this approach requires extra communication between the software agents for the dynamic peer-to-peer coordination experimental results validate the potential of the NPI paradigm (see Chapter 7).

At the intra-domain pricing level one of the main challenges is the evaluation of the marginal cost of providing services. The opportunity cost of using specific network resources should dynamically reflect the probability of blocking future incoming demands when using a certain amount of network capacity, i.e., bandwidth on a link. In current communication networks, however, costs are mainly considered as constant amounts independently on the current network utilisation. Several works have addressed the need of dynamically evaluating the costs in such a way that the prices efficiently reflect current network utilisation versus pre-fixed demands blocking probabilities. As identified in [342], the two components of the opportunity cost are the "shadow price for reserving one unit of bandwidth" and the "shadow price for using one unit of bandwidth". However, it is not clear how these factors relate to the current bandwidth resource usage. In [80], Fankhauser et al. define a base pricing scheme that considers an over-booking ratio fixed by every provider and the current available bandwidth in order to compute a traffic factor which takes into account the network congestion degree. However, this traffic factor intervenes in the price computation only if the network is congested. It is assumed indeed that there is revenue only if there is congestion. Therefore, under the congestion threshold, there is no way of correlating the resources utilisation degree with prices. This strongly limits the possibility of optimising the allocation of network resources.

The main idea beyond the pricing model developed in the NPI context is to dynamically compute costs and thereby prices by considering the current network resource criticalness, which can be quickly estimated by making use of the Diff-BIH decomposition of the network (see Section 3.4.8). The lower a blocking island is in the Diff-BIH, the less critical are the links clustered in it. Therefore, by using resources in a lower BI, a better bandwidth connectivity level is maintained. This has the main effect of reducing the risk of future demand allocation failures. Notice that the intra-domain pricing scheme adopted by every NPA is not mandatory for the NPI approach to work.

5.2.1   The Intra-Domain Pricing Approach

The intra-domain pricing scheme adopted by every provider in the NPI context relies upon the following main assumptions:

Assumption 5.2.1. There are two main components that intervene in the computation of the price that a provider fixes for the allocation of a specific service demand: a base price fee that depends on the specific service requirements and a networking price fee that depends on the current network resources state.

The formal definitions of *base price* and *networking price* are given in Section 5.2.1.

Assumption 5.2.2. The price of a link (and therefore of a service) is assumed to be fixed for the overall duration of the service set-up process.

This assumption simplifies the analysis of our techniques. However, it would be possible to relax it and consider how, during the service consumption from users, prices could be dynamically re-negotiated between providers and users (i.e., dynamic service re-negotiation). Since this work focuses on the service setup phase, the dynamic service re-negotiation is considered as a possible future extension of the NPI approach (see Section 9.3.2).

Assumption 5.2.3. The resources costs inside every network are computed on-demand, based on the criticalness factor of every link used for allocating the demand.

Since every NPA maintains the Diff-BIH decomposition of network resource availability (that is used also during the node consistency phase of the DAC process, see Section 4.3.2), the computation of the criticalness factor is computationally very cheap (see Section 3.4.2).

Assumption 5.2.4. The on-demand computed cost of a link is assumed to be fixed for the overall duration of the service set-up process.

Similarly to what happens for the price, it would be possible to relax this assumption and consider dynamic service re-negotiation.

Assumption 5.2.5. The information needed for the base price computation is contained in the service demand specification that every provider involved in the MuSS process receives from the initiator.

Assumption 5.2.6. The information needed for the networking price and costs computation is retrieved by the Diff-BIH decomposition of the network that every provider maintains.

The Diff-BIH decomposition makes it possible to quickly perform the node consistency check (see Section 4.3.2) and dynamically compute the criticalness factor of the network links (see Section 3.4.4).

Assumption 5.2.7. ¿From the provider point of view the main objectives when pricing resources are:

- Reducing prices for increasing the probability that end users will accept the offer. This aims at maximising the end user satisfaction and being competitive with respect to other providers in the market.

- Minimising costs for maximising the provider benefit while minimising the prices. This means that among the available resources the least expensive should be chosen first.

The NPI Pricing Function

There are two main components that intervene in the computation of the price that a provider asks for the allocation of a specific service demand: a *base price fee* that depends on the specific service requirements (i.e., QoS characteristics and duration), and a *networking price fee* that depends on the current network resources state (see Assumption 5.2.1). Therefore, when a provider prepares a proposal (or offer) for the allocation of a specific service demand d these two components are computed as follows.

Definition 5.2.8 (Base Fee). Given a service demand $d = (x, y, qos, dur)$, with x and y the source and the destination node, qos the vector expressing the minimal required bandwidth β^r, the required end-to-end delay e^r, and the required service level l^r, and dur indicating the duration of the service, the *base fee* $b(d)$ of allocating d is determined by the product of a constant base factor κ and the traffic volume v generated by the allocation of d:

$$b(d) = \kappa v$$

where $\kappa \in \{\kappa_G, \kappa_S, \kappa_B, \kappa_P\}$, with $\kappa_G > \kappa_S > \kappa_B > \kappa_P$, and $v = \beta^r dur$.

This price component depends on the specified end user requirements (i.e., usage-based fee) and gives therefore an incentive to the users to declare the true QoS needed for the allocation of a specific service. By requesting higher QoS guarantees users would get offered a higher price. On the other hand, lower QoS requirements would not allow users to obtain the required quality of service. Base prices for different service levels (i.e., κ values) have to be fixed quantitatively. Since this depends on the user behaviour and on the network configuration, these base prices have to be studied by simulation and by observation of existing networks [80].

Definition 5.2.9 (Networking Fee). Given a service demand d and its feasible set F, the networking fee $n(d)$ that a provider fixes for allocating d is given by the sum of the individual networking fees $n(l_j)$ of all the links l_j forming the intra-domain route $r \in F$ selected to allocate d:

$$n(d) = \sum_{l_j \in r} n(l_j)$$

where n is a pricing function, which assigns a real positive value to each link l_j depending on its available bandwidth $\beta(l_j)$ and its criticalness factor $\mu(l_j)$.

This component is a network *state-based* fee since it takes into account the current availability of network resources (i.e., bandwidth). At the same time, because of the dependency on the criticalness factor, the networking fee represents a *cost-based* fee. The criticalness factor is indeed directly related to the cost of the network resources used. The networking price component is equivalent , for instance, to the fee that in [80] has been called the "traffic factor", since it actually estimates the impact in terms of produced traffic on resources that have different levels of criticalness.

In the NPI context, the main advantage of introducing the networking fee component into the computation of the service price is the possibility to control and reduce network congestion. Users may be willing to refine their QoS requirements so that they can get lower prices, since less critical links (i.e., lower μ and therefore lower cost) can be used in the network. The specific pricing function n adopted in the NPI context is:

$$n(l_j) = \mu(l_j) \exp^{-q\beta(l_j)} \tag{5.3}$$

where $\mu(l_j)$ is the criticalness factor of link l_j, q a positive constant value and $\beta(l_j)$ the available bandwidth on link l_j. This pricing function makes it possible to increase prices when the criticalness of the link increases and when the residual available bandwidth on the link decreases [341], [80], [337].

Finally, the price offered to the potential customer for the allocation of a specific service demand is computed as a combination of the base price and networking fee components as follows.

Definition 5.2.10 (Service Demand Allocation Price). Given a service demand d and its feasible set F, the service demand allocation price $p(d)$ is given by the product of the base fee and the networking fee of d:

$$p(d) = b(d)n(d)$$

The Availability Criticalness Evaluation Cost Function

As anticipated in Section 3.2.2, given a provider network graph $\mathcal{G}_I = (\mathcal{N}_I, \mathcal{L}_I)$ modelling a network I, every link $l_i \in \mathcal{L}_I$ is associated with a cost $c(l_i)$. This parameter estimates the cost of allocating network resources from the provider (or firm) perspective. In order to compute the cost $c(l_i)$ the *discrete* cost function $c : \mathcal{B}^o \to \mathbf{R}$, which assigns a real positive value to each bandwidth level offered on a link $\beta(l_i) \in \mathcal{B}^o$, needs to be identified. In the NPI context, the finite set of offered bandwidth levels is given by $\mathcal{B}^o$ (see Definition 3.9). In principle, there may be infinitely many discrete bandwidth levels corresponding to the different values of available bandwidth on the physical links. However, in this volume, since a discrete service model is assumed, the link cost $c(l_i)$ can take up to $b^o = |\mathcal{B}^o|$ finite, different values [9].

Definition 5.2.11 (Non-increasing cost function). A cost function c is *non-increasing* if for any two links l_i and l_x with two different bandwidth values β_i and β_x identifying their respective father blocking islands, where $\beta_i > \beta_x$ and $i \neq x$, it holds:

$$c(l_i) \leq c(l_x)$$

Definition 5.2.12 (Availability Criticalness Evaluation cost function). Given a network I and its domain graph $\mathcal{G}_I = (\mathcal{N}_I, \mathcal{L}_I)$, the *Availability Criticalness Evaluation* cost function $c(l) : \mathcal{B}^o \to \mathbf{R}$ is a discrete non-increasing cost function associating a real positive number to each parameter value μ_l that expresses the criticalness of a given link $l \in \mathcal{L}$, i.e.,

$$c(l) = k\gamma(\mu(l))f(\beta^r)$$

with k a positive constant value, $\gamma(\mu(l))$ a non-increasing function of the criticalness factor $\mu(l)$ of the link l and $f(\beta^r)$ an increasing function of the required bandwidth β^r. $c(l)$ is indicated as the *opportunity cost* of allocating the link l.

---

[9]In [248], Raz and Shavitt present an approach for evaluating the cost function in Differentiated Service frameworks very similar to the one presented in this volume. However, their focus in on "optimal partition of QoS requirements" for routing rather than for pricing issues.

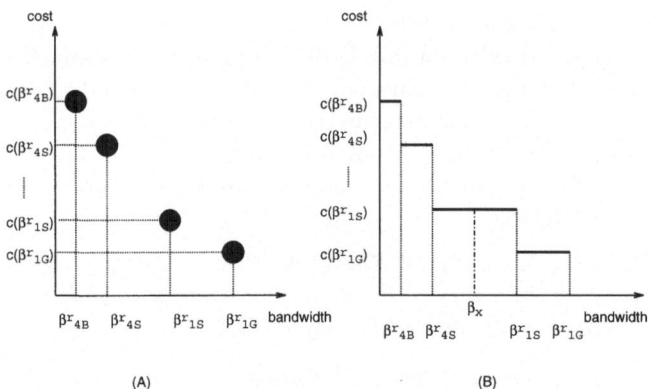

Figure 5.6: *Case (A): The discrete cost function associates a cost for every distinct possible bandwidth level $\beta_{jk}^r \in \mathcal{B}^o$. Case (B) shows its representation as a step function. The opportunity cost of a link with available bandwidth β_x and father β_{1S}^r-BI is determined by the cost of the bandwidth level β_{1S}^r.*

Since the available bandwidth β_F of the father BI (see Definition 3.4.3) of a link l_i directly determines the criticalness factor $\mu(l_i)$, to adopt the ACE policy means to approximate the cost of a link l_i with available bandwidth $\beta(l_i)$ with the cost of l_i as if its available bandwidth was $\beta_F \in \mathcal{B}^o$.

Remark that when considering two distinct links l_1 and l_2 with $\beta(l_1) > \beta(l_2)$ it is possible that the link l_2 is less critical than the link l_1. This happens when the father BI of l_2 is lower in the Diff-BIH than the father BI of the link l_1. This is due to the fact that the opportunity cost of a link is not directly related to the amount of its available bandwidth, but rather to its criticalness in the Diff-BIH. The criticalness factor estimates the risk of isolating some network nodes in terms of reachability when allocating the resources on a particular link.

A suitable way to visualise the cost function is to consider a step function like the one Figure 5.6 case (B) displays. For any link l_x, with available bandwidth β_x, its opportunity cost $c(l_x)$ is inversely related to the available bandwidth of its father BI. In this example, the father BI of l_x is β_{1S}^r. The cost function is defined only for discrete points $\beta_{jk}^r \in \mathcal{B}^o$, and it is easy to see that an optimal allocation is always at these points, since sliding rightwards on a step increases the amount of bandwidth used/offered without increasing the cost. Two main cost functions families have been considered:

- $c(l) = k\mu(l)\beta^r$, with k a constant positive value.

- $c(l) = k\exp^{k'\mu(l)}\beta^r$, with k and k' constant positive values.

Both functions (as displayed in Figure 5.7) are decreasing when the amount of available bandwidth of the father BI of a link is increasing. Depending on the

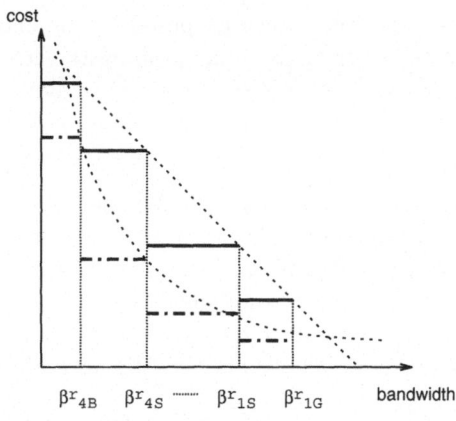

Figure 5.7: *In this figure two different cost functions are displayed. Both their step approximations are decreasing when the amount of available bandwidth of the father BI of a link is increasing. In this example, for the same amount of available bandwidth $\beta^r_{jk} \in \mathcal{B}^\circ$ the cost of a link is higher when considering the linear cost function. Therefore, the choice of a different cost function can have a direct impact on providers profits.*

selected function the same criticalness level can have different costs. In the experiments discussed in Chapter 7, the first type of cost function listed above has been used. How this has a direct impact on the providers profit computation is discussed in more details in Section 8.2.2.

To conclude, the opportunity cost of allocating a service demand by considering all the network resources used for satisfying the demand, can be formally express as follows:

Definition 5.2.13 (Opportunity Cost of Allocating a Service Demand). Given a service demand d and its feasible set F, the opportunity cost $c_i(d)$ that a provider i encounters for allocating d is given by the sum of the opportunity costs of all the links l_j forming the intra-domain route $r \in F$ selected to allocate d:

$$c_i(d) = \sum_{l_j \in r} c(l_j)$$

In the NPI context, provider agents can make use of two main alternative pricing policies.

- The *ACE policy* makes use of the ACE cost function (as expressed by Definition 5.2.12) in order to dynamically compute the opportunity cost of using specific links, taking into account their current BI criticalness factor.

Moreover, also the prices proposed to the end users are correlated to the criticalness factor of the links forming the selected intra-domain route (see Equation 5.3). In this way, the profit, computed as the difference between prices and costs, is directly proportional to the BI criticalness of the resources used.

- With the *No Criticalness* (NOC) policy, costs are assumed to be constant values that every provider statically establish[10]. Service demand allocation prices are computed also in this case as the product between a base price component and a fee component depending on the current network state. While for the computation of the base price agents make use of the same expression given in Definition 5.2.8, the networking fee is computed as follows:

$$n(l) = p \exp^{-q\beta(l)} \tag{5.4}$$

with p and q constant positive values. With this latter approach, the network state component is not related to the BI decomposition of the network, but simply to the amount of available bandwidth $\beta(l)$ on any link l used. Again this corresponds to traditional approaches [342], [341].

The average normal utility, profit and social welfare are higher in all the simulated scenarios when the provider agents dynamically compute prices and costs (i.e., ACE policy). In Section 7.2.4 and 7.3.3 an experimental analysis allows to better compare the different performance of providers making use of these two distinct policies.

---

[10]This is actually the kind of approach that network providers in real environments adopt.

Chapter 6

Alternative Approaches for Providers Coordination

*My decided preference is for the fourth
and last of these alternatives.*
– GLADSTONE,

¿From the end user perspective, today's networks offer in principle two main ways of requiring the allocation of a service demand spanning several domains [66].

- The *single-provider* approach: The end user contacts a specific provider that is responsible for setting up the necessary interconnections with the distinct network providers involved in the service allocation. In this case, the coordination of different domains is resolved at the level of peer providers and the end user stipulates a unique contract with the contacted operator.

- The *multiple providers* approach: The end user (or a service provider acting on behalf of him) contacts multiple network providers from the source to the destination domain. Therefore, several negotiations with distinct operators have to take place and different contracts may need to be defined.

This second approach however is not extensively adopted. Nowadays, end users are more and more willing to interact with a single provider that focuses on the required relationships taking care of resolving all the interworking issues involved in the service provisioning [66], [155]. Within the single-provider contact paradigm, from the network operator perspective, the coordination of distinct domains can be achieved in different ways. Therefore, in this volume the *single-provider* approach has been explored. In this chapter, one of the main objectives is to present and discuss several possible mechanisms for the coordination of peer providers. The central idea is to underline weaknesses and strengths of alternative solutions with respect to the NPI approach described in Chapters 4 and 5.

Two main alternative procedures have been considered and experimentally analysed.

- The *Fixed agreements Solution* (FAS) is a decentralised approach in which peer-to-peer coordination is regulated by pre-existing static agreements. The provider contacted by the end user is responsible for formulating a unique offer by interacting with other operators on the basis of prices and traffic profiles fixed beforehand, see Section 6.2.

- The *COalition Based* (COB) approach relies upon the dynamic coordination of providers forming on-demand coalitions and agreements supervised by a central controller. Within each coalition, this supervisor is responsible for collecting information from different providers and formulating a unique offer to the end user, see Section 6.3.

A formal analysis of the relation between the NPI approach, which is based on individual bargaining agents controlling distinct domains, and the COB paradigm, in which a central entity controls the whole multi-provider network, is given in the second part of the chapter (Section 6.4). The main purpose is to find out if and how the criticalness factor of a given link may vary when passing from a distributed control structure to a centrally supervised scheme. The relative value of the network resource criticalness parameter has a major impact on the solution offered to the end users in terms of QoS, costs and prices. Therefore, for the same network scenario and the same resource state, the final outcome that the NPI and the COB procedures determine may differ. Finally, in order to conclude this overview of alternative approaches for the coordination of distinct providers, a short survey of the main NPI characteristics is given.

6.1   A Fully Centralised Approach

Given the allocation of a certain service demand, a very intuitive way of determining the end-to-end Pareto [333] optimal solutions[1], either in terms of end-to-end QoS characteristics or in terms of provider profit or social welfare, is to delegate the decision process to a central entity with a detailed view of all possible networks in the multi-provider scenario.

The central providers global coordinator, based on full knowledge of all individual provider networks, would be able to determine the Pareto optimal allocation of a given service demand by performing an exhaustive search of the solution

---

[1]A route is Pareto optimal if improving the value for one metric cannot be achieved without making worse the value for at least one of the other metrics.

space. If the goal is to determine the Pareto optimal end-to-end route in terms of bandwidth used (i.e., least bandwidth), or in terms of cost or delay (i.e., either least cost or least delay), this central providers global coordinator could make use of a unicast routing algorithm such as the Dijkstra [67] or the Bellman-Ford [16] algorithms in polynomial time complexity. As discussed in Section 4.3.2, although whenever multiple QoS constraints are considered, routing a single demand becomes in general intractable, the *bandwidth-constrained delay-optimisation* routing process can as well be solved in polynomial time. In addition to the possibility of exploring the whole solutions space and thereby determining the Pareto optimal routes, within this kind of centralised approach an impartial and fair process would be easier to implement and control than in the case of distributed and autonomous entities.

Nevertheless, such a centralised approach it is not feasible in real networks for a number of reasons.

- *Lack of scalability*: the presence of a central entity responsible for determining the Pareto optimal routes may require a massive transfer of data from every distinct network. Hence, when the size and the number of domains involved increase this data exchange could heavily slow down the whole allocation process.

- *Heavy time consumption*: centralised decisions and actions can be very time-consuming. This is due to the additional time it takes for the information describing every distinct network to flow to the providers leader, the additional time needed to create a global coherent vision out of partial visions, and the additional time it takes for the controller to give a feedback to every single provider.

- *Lack of robustness*: if the central coordinator breaks down the whole service set up process can fail.

- *Lack of confidentiality*: to communicate strategic information to the central controller requires mechanisms guaranteeing a secure exchange.

For overcoming these main limitations, the common trend in network management and control is to adopt decentralised solutions (see [196], [154] and several works in [124]). This trend is even stronger in multi-provider environments where the geographical distribution of resources, the scalability needs and the confidentiality of strategic information force the adoption of decentralised approaches. As discussed in Section 2.3, the existing interworking schemes heavily rely upon the supervision of human operators. At the intra-domain level, these operators control low level network components through the usage of tools and/or application programming interfaces. At the inter-domain level, they directly supervise low level interfaces (e.g., TMN-X interface) and usually interact between each other via telephone, facsimile, e-mail, etc. This makes the overall interworking task very slow (for instance, several weeks can pass before effective inter-domain network configuration changes take place) and quite inefficient. Considering these aspects, what seems more suitable for future scenarios, is a decentralised and dynamic approach based on static and/or mobile software entities collecting network state information and which have the ability to directly interact with each other and invoke effective changes in network components, without the interaction of a human operator.

6.2   The Fixed Agreements-based Solution

The Fixed Agreements Solution (FAS) is a distributed and automated approach in which coordination of distinct agents is regulated by static long-term service level agreements. Despite the lack of negotiation, the automation of many control and management tasks makes it possible to accelerate the set up of an end-to-end route crossing several domains. This kind of architecture represents the closest mechanism to the current human-driven approach for multi-provider interactions. Today, every network operator has its set of transport providers to reach other domains in other countries or continents, either by cable or by satellite. Quite static contracts establish how the different networks can connect to each other. Whenever changes in the inter-domain arrangements are required, human controllers directly supervise low level interfaces and usually interact between each other in order to modify the existing configuration and SLAs.

The main idea behind the FAS approach is to facilitate the interworking process by introducing software agents that automate the providers coordination process based on the existing contracts. There is no dynamic negotiation since binding agreements already establish prices and minimal service levels that providers are expected to offer to each other. This kind of procedure is very similar to the bandwidth broker based approach proposed by Clark and Fang in [51]. In their work, at every network access point actual traffic is metered and tagged according to the conformance or not to the pre-fixed contracts regulating multi-domain interactions. Their model charges only traffic that causes congestion (other traffic is covered by flat-rate charges).

In the FAS context, pricing is regulated by the ACE policy (see Section 5.2.1). Human operators interacting with their agents are responsible for eventually modifying the existing peer-to-peer SLAs. Therefore, the main weakness of the FAS approach is that it is not possible for distinct providers to dynamically re-negotiate the traffic profiles based on the current network state. While the FAS procedure benefits from the main strengths of decentralised paradigms (see Section 6.5), in comparison with the NPI solution, the most fundamental limitation is the poor flexibility of static inter-domain agreements to changing networking conditions. Since it is not possible to dynamically accommodate costs and prices for specific service levels, the optimisation of network resource utilisation is strongly limited. This increases the probability of rejecting a higher number of service demands determining losses in terms of providers profit and end users satisfaction. Experimental results presented in Section 7.4 confirm the poor flexibility of the FAS approach. In comparison to the NPI approach, it is possible to observe, for instance, a significant degradation in terms of the average number of service demands succesfully allocated. This directly impacts both the end users utility and the providers profit.

6.3 The Coalition-based Approach

The *COalition Based* (COB) scheme relies upon the dynamic formation of providers coalitions supervised by a meta-controller entity, see Figure 6.1. The fundamental idea is that groups of network domains accept to join each other to better coordinate their interactions by delegating part of the allocation process for a specific demand to a meta-NPA. Every individual coalition is controlled by a central providers leader (or meta-NPA), which has the role of collecting information from every network to be able to supervise the coordination of distinct providers. The various agents belonging to the same coalition have in this case some general long-term acknowledgements about each other roles, interests and responsibilities during the service set up process.

The COB paradigm relies upon the dynamic coordination of a group of NPAs involved in the allocation of a specific service demand by delegating the control of the final phase of the MuSS process to a central *Coalition Leader Agent* (CLA). This final phase corresponds to the global offer making step that, in the NPI paradigm, is solved by bi-lateral negotiations taking place between the *initiator* NPA and the other NPAs along the abstract path selected to allocate the service demand (see step 5 of the DAC mechanism, Section 4.3.2). Within the COB approach, when distinct NPAs successfully terminate, in a distributed fashion, the arc consistency task, i.e., they have non-empty feasible sets for the demand to be allocated (see Section 4.3.2), they delegate to the CLA the definition of a specific global offer for the end user. This offer consists of a unique end-to-end route with certain QoS characteristics at a global price, which includes fees for every selected route in every network. In order to formally define the COB procedure, the following preliminary concepts are given.

Definition 6.3.1 (Providers Coalition). Given a provider set $\mathcal{P}$ and a specific multi-provider service demand d to be allocated, a providers coalition $\mathcal{C}_x(d)$ consists of a group of network provider agents in $\mathcal{P}$ that cooperate for the allocation of d under the supervision of a coalition leader agent CLA_x.

Definition 6.3.2 (Coalition Leader Agent). Given a specific providers coalition $\mathcal{C}_x(d)$, the coalition leader agent CLA_x is the entity responsible for collecting individual feasible sets from every NPA $i \in \mathcal{C}_x(d)$, in order to compute a coalition global offer o_x for the end user requesting the allocation of d.

For every given coalition, it is assumed that there exists a unique coalition leader that aims to increase the social benefit of the coalition, which is given by

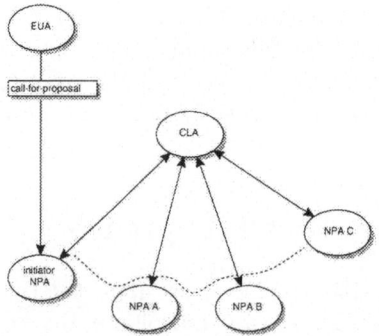

Figure 6.1: *Within the COB approach the coordination of distinct providers is supervised, during the final phase of the MuSS process, by a central Coalition Leader Agent (CLA).*

the sum of the providers profits, while choosing the less critical routes in every domain (see the Definition 3.4.10 for the concept of *critical route*).

Definition 6.3.3 (Coalition Global Offer). Given a service demand d and a providers coalition $C_x(d)$ controlled by the coalition leader agent CLA_x, the global offer o_x that the CLA_x proposes to the end user for the allocation of d is called the *coalition global offer*. This consists of a specific end-to-end route r_x with specific QoS characteristics at a certain global price that includes fees for all individual intra-domain routes forming r_x.

The QoS characteristics of the route r_x are the available bandwidth and the global end-to-end delay. The global price corresponds to the sum of all the revenues that the CLA will redistributed among the NPAs in the coalition. Therefore, for every successful negotiation between the end user and the CLA, the global price (i.e., the amount of money) that the end user accepts to pay is the *coalition revenue* (CR).

6.3.1   Coalition Formation

The process enabling the formation of a certain coalition starts when the end user submits a call for proposal for the allocation of a specific service demand d. The contacted *initiator* NPA computes on-demand a list P^D of abstract paths satisfying bandwidth requirements and minimising the estimated end-to-end delay, as discussed in Section 4.3.2. The set P^D consists of all the possible alternative coalitions for the allocation of d. At this point, there are two main distinct approaches:

- *One Coalition.* Over the possible alternative coalitions in P^D, the *initiator* NPA selects a specific abstract path $\mathcal{A}_x$ by applying the heuristics listed in Section 4.3.2. A unique coalition $C_x(d)$ formed by the providers along $\mathcal{A}_x$ is created. In the NPI system, a new agent is generated for assuming the role of

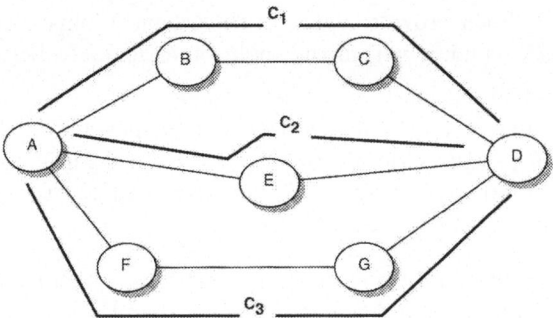

Figure 6.2: *In this example, there are three main abstract paths between the two different networks A and D. For each path a distinct providers coalition exists: C_1, C_2 and C_3. The three coalitions may search in parallel for the Pareto optimal solutions along every distinct path. Then, a meta-CLA would select the optimal route over all the possible alternatives.*

the coalition leader CLA_x (in real environments coalition leaders are usually third trusted parties). The CLA_x is responsible for defining the global offer for the end user and redistributing the revenue to every NPA in the coalition. If for some reason the service allocation within the coalition $C_x(d)$ fails, it is possible to investigate the allocation of d by creating a new coalition $C_y(d)$ corresponding to a new abstract path $\mathcal{A}_y \in P^D$, with $x \neq y$.

- *Multiple Coalitions.* For every abstract path $\mathcal{A}_j \in P^D$ a distinct coalition $C_j(d)$ is formed and a distinct coalition leader agent CLA_i is created for each of them, see Figure 6.2. However, in this way some provider agents, at least the source and the destination domains, may belong to more than one coalition. Since each NPA have to finally commit for a single coalition (usually the one maximising its own profit), it may happen that there is no unique coalition, which all NPAs agree upon. It would be therefore necessary to delegate the choice of a unique coalition among all the possible ones to a meta-CLA supervising all the different groups. This meta-CLA may decide, for instance, to select the coalition maximising the revenue.

In this context, we investigate, as a first step, the *one coalition* approach. The multiple coalitions paradigm could be considered as a future extension of this simpler approach.

In order to make the creation of mutually beneficial providers coalitions possible, the following assumptions are made.

Assumption 6.3.4. Distinct provider agents can communicate and coordinate each other activity by choosing among various known interaction protocols.

Assumption 6.3.5. Distinct provider agents can supply services to each other and for this purpose they can execute side-payments.

Assumption 6.3.6. Each provider agent in the scenario joins a coalition only if it can benefit at least as much within the coalition as it might benefit by itself.

When individual self-motivated agents decide to cooperate, the reason is that by collaborating they increase their own benefit (or payoff). In the COB context, therefore, distinct providers decide to join a certain coalition $C_x(d)$ only if the profit they derive by being part of $C_x(d)$ is greater than, or at least equal to, what they can obtain by staying outside the coalition.

- For a specific provider to be part of a coalition, which is responsible for allocating a given service demand d, means that there is potential profit to be made by allocating a certain amount of network resources. The providers benefit is evaluated as the difference between the revenue and the costs a provider incurs for the service demand allocation. The revenue for every provider is established by the coalition leader agent by redistributing the CR. The cost includes the internal expenses incurred for the resources that need to be allocated, the computational and the communication costs. Hence, assuming that providers do not sell at a loss (see Assumption 5.1.3), a provider j accepts to be part of the coalition only if the portion of the CR that the CLA redistributes to j is greater than, or at least equal to, the totality of costs that the provider faces for allocating d.

- Not being part at all of the coalition means for a provider not to be able to obtain any profit out of that specific service demand allocation. Nevertheless, the refusal to participate to the coalition it is not necessarily equivalent to decrease the potential provider benefit: the network resources left free may be used in the future for more profitable deals. In this case, providers should speculate about the future incoming traffic. It is however very difficult (and out of the scope of this work) to make effective predictions about multi-provider traffic profiles given that providers are not willing to reveal this kind of information.

Proposition 6.3.7 (Necessary Condition for a Successful Coalition). *Given the allocation of a service demand d and the coalition $C_x(d)$ corresponding to the group of NPAs along the selected abstract path A_x, a necessary condition for the the service demand allocation along the abstract path A_x to succeed is that the coalition revenue ensures a positive benefit for every NPA $\in C_x(d)$.*

PROOF: Every NPA j joining $C_x(d)$ incur a specific cost c_j given by the sum of internal resources costs, communication and computational costs.

Therefore, if the portion of the coalition revenue CR_x that each NPA j along A_x receives is not greater than, or at least equal to, c_j, then the NPA j has incentives to abandon $C_x(d)$ because of the main assumption of individual rationality of provider agents (see Assumption 5.1.5). Thereby, the allocation along A_x fails.

□

Corollary 6.3.8 (Minimal CR). *Given the allocation of a service demand d_k and the coalition $\mathcal{C}_x(d)$ corresponding to the group of NPAs along the selected abstract path $\mathcal{A}_x$, the minimal CR_x guaranteeing the stability of the coalition is greater than, or at least equal to, the sum of costs every NPA incurs being part of the coalition $\mathcal{C}_x(d)$.*

PROOF:
 The profit Π_i of every NPA $i \in \mathcal{C}_x(d)$ is given by the provider revenue p_i less the cost c_i. Since every provider revenue is a portion of the global CR_x, by directly applying Proposition 6.3.7, it follows that for the success of the coalition the following condition must hold:

$$CR_x = \sum_{i \in \mathcal{C}_x(d)} p_i \geq \sum_{i \in \mathcal{C}_x(d)} c_i$$

□

Proposition 6.3.9 (Grand Coalition Dominance). *Given the set up process of a service demand d and a selected abstract path $\mathcal{A}_x$ for allocating d, the coalition maximising the social welfare for any NPA i along $\mathcal{A}_x$ is the grand coalition $\mathcal{C}_x(d)$ including all NPAs $\in \mathcal{A}_x$.*

PROOF: All NPAs along $\mathcal{A}_x$ are best off by forming the *grand coalition* where all agents operate together, since if one of them drops $\mathcal{C}_x(d)$ the allocation of the service demand along $\mathcal{A}_x$ fails and therefore the total CR_x is zero as well as the profit for every NPA along $\mathcal{C}_x(d)$. □

 This proposition proves that within a selected abstract path the multi-provider service allocation is a *super-additive* game. This game theoretical notion means that any pair of two coalitions formed by two distinct groups of agents along the selected abstract path are best off by merging into one coalition [283]. In the multi-provider service set up process this is quite intuitive, since disjoint coalitions of providers along the same abstract path are not able to determine a complete end-to-end route. On the other hand, without making any a priori choice about which abstract path to follow, the MuSS framework is *not a super-additive* environment. For provider agents belonging to disjoint coalitions along alternative paths it is not beneficial to form one unique grand coalition. The main reason is that the CR would have to be redistributed among a higher number of participants reducing therefore the profit for individual providers.

6.3.2 Global Offer Computation in a Providers Coalition

The COB approach can be considered as a combination of a decentralised NPI-like paradigm with a centralised meta-control level that the coalition leader entity exerts in the final phase of the service set up process. This is done with the purpose of maintaining on one side the major advantages of a distributed approach and, on the other side, gaining the benefits of a centralised controller. This neutral *supra-partes* entity, in alternative to bi-lateral negotiations, has the potential to define an end-to-end offer that can accommodate the end user request and maximise the social benefit for the coalition guaranteeing a higher degree of fairness and a better load balancing among all involved providers involved. For this purpose, the CLA has to collect information describing which intra-domain routes are available in distinct domains.

Given the allocation of a certain service demand d, every NPA i belonging to the formed coalition $C_x(d)$ individually verifies the set of possible local routes that are compatible with the existing intra- and inter-domain constraints. At the end of the arc consistency phase (i.e., step 4 of the DAC mechanism, see Section 4.3.2), every NPA i submits its *feasible set* F_i[2] to the CLA$_x$ responsible for $C_x(d)$. At this point, there is the main differentiation between the DAC approach proposed in Chapter 4 and the COB paradigm. Instead of peer-to-peer negotiations between the *initiator* NPA and the other NPAs along the selected abstract path (i.e., step 5 of the DAC mechanism), the global offer computation is done in this latter case by the CLA$_x$.

In the COB context, given the allocation of a service demand d along the abstract path $\mathcal{A}_x$, the global offer computation is performed as follows.

- For every NPA $i \in C_x(d)$, the CLA$_x$ collects the feasible set F_i.

- Given all the feasible sets, the CLA$_x$ determines the global offer o_x by considering the end-to-end route r_x satisfying the service demand QoS requirements and combining the least critical routes from every distinct domain along the abstract path $\mathcal{A}_x$.

- The global price p_x is computed as the sum of the prices p_i of the individual route fragments r_i forming the end-to-end route r_x. Every p_i is computed by applying the formula given in Definition 5.1.9.

- The CLA$_x$ submits the global offer o_x to the end user:

    - If the end user accepts o_x, the CLA$_x$ redistributes the CR to every NPAs in the coalition. The portion of revenue CR$_i$ that every NPA i gets is directly proportional to the criticalness of the specific selected intra-domain route r_i. More precisely, the CR$_i$ corresponds to the computed price p_i per route fragment.

    - If the end user rejects the offer, an alternative proposal can be made by considering other available intra-domain routes in the feasible sets. If no other global offers can be made the service allocation along $\mathcal{A}_x$ fails. A new coalition $C_y(d)$ can be investigated, if the *initiator* NPA

---

[2]See Definition 4.3.3 for the concept of *feasible set*.

activate the search along another available abstract path $\mathcal{A}_y$, otherwise the service allocation fails.

The redistribution of the coalition revenue is a fundamental task on which the stability of the whole coalition depends. More precisely, the social benefit should be redistributed in a way that every agent (or subgroup of agents) is better off staying in the social welfare maximising coalition structure than by separating into a new coalition [153], [329]. In the COB context, the necessary condition for a NPA j to accept to be part of a coalition $\mathcal{C}_x(d)$ is to get a revenue that is greater than, or equal to, the provider costs encountered for supporting the specific service demand d because of the individual rationality assumption. By dropping $\mathcal{C}_x(d)$, the NPA j does not have any guarantee that an alternative selected abstract path $\mathcal{A}_y$ will include its domain. In that case, the risk is to leave a coalition that ensures a certain profit for a situation in which the profit might be zero. Nevertheless, further investigation would be needed in order to establish the sufficient conditions for the COB approach to guarantee the stability of the coalition.

6.3.3 The Clarke Tax Mechanism for Providers Coalitions

Without any further control, the coalition revenue redistribution process discussed in the previous sections does not give any guarantee that the dominant strategy of individual NPAs in the coalition is to reveal the truth. A strategy is said to be *dominant* if an agent is best off by using that specific strategy no matter what strategies the other agents use [166].

Since in the COB context the profit every NPA gets out of a successful service allocation is directly proportional to the declared cost (and thereby price) of every route in the feasible set, each NPA has incentives in declaring higher costs (i.e., requiring higher revenues). One could argue that by increasing the costs and therefore the required revenue, the probability that the end user will reject the global offer increases, since the CLA has to ask for a higher global price. For this reason, even without further control, malicious agents should carefully consider the possibility of arbitrarily increasing the declared costs. However, in order to ensure that *true telling* is dominant, it is necessary to define and add specific control techniques. This can be done in the COB context by imposing the Clarke tax mechanism.

The Clarke tax [52] is a classical technique for ensuring that agents reveal their true preferences. The main idea is that each agent pays a tax corresponding to the portion of its bid (i.e., intra-domain route declared profit) that makes a difference to the final service allocation outcome (i.e., to the global coalition revenue).

To better understand this mechanism, consider the following example. For the allocation of a given service demand d, the coalition $\mathcal{C}_x(d)$ consists of three distinct provider agents: NPA A, NPA B and NPA C. Given the feasible set of every NPA and their declared costs and prices, the CLA$_x$ is able to formulate three distinct end-to-end routes: r_1, r_2 and r_3. Table 6.1 shows the different parameters necessary to compute the tax every NPA has to pay. An asterisk is used to mark the selected route in each situation. When all the agents reveal their true preferences - influencing thereby the CLA choice - the end-to-end route r_2 is selected, since it gives the higher coalition profit $\Pi = 18$. If NPA A was not contributing to the selection, the route r_3 would be selected instead of r_2 and the difference in terms of profit would be $\Pi'(r_3) - \Pi'(r_2) = 12 - 11 = 1$. Therefore, the NPA A affects the final outcome by a "magnitude" of 1. This represents the value of the tax the CLA_x imposes on NPA A. Following the same procedure, it is possible to compute the taxes for NPA B and NPA C. Notice that NPA B is not fined because even if it had not revealed its profit, the end-to-end route r_2 would still have been selected.

Agent i	$\Pi_i(r_1)$	$\Pi_i(r_2)$	$\Pi_i(r_3)$	$\Pi'(r_1)$	$\Pi'(r_2)$	$\Pi'(r_3)$	Tax for i
NPA A	6	7	2	9	11	12*	1
NPA B	8	2	4	7	16*	10	0
NPA C	1	9	8	14*	9	6	5
tot. Π	15	18*	14	–	–	–	–

Table 6.1: *The Clarke tax computation. For every end-to-end route, the 2nd, 3rd and 4th column display the declared profits for every provider for the different end-to-end routes r_1, r_2 and r_3. Then, the following columns show the total coalition profit Π' along the different routes, if the declared profit of a specific agent had been zero. An asterisk marks the winning choice in each different situation.*

Consider now the case in which the NPA A is a malicious agent that reveals a higher expected profit, e.g., 7 instead of 6, for the end-to-end route r_1 and artificially decreases the profit for the end-to-end route r_2, e.g., 2 instead of 7, see Table 6.2. This could be done with the purpose of getting a higher portion of the CR or increasing the probability that the coalition leader will prefer r_1 over the other possibilities. In this example, the first consequence is indeed that the end-to-end route r_1 is selected, since it gives the highest coalition profit $\Pi = 16$. At the same time, another main effect is that the taxes values change. In particular, the malicious NPA A has to pay a tax equal to 3. Therefore, while by revealing the truth the net benefit is $\Pi_A = 7 - 1 = 6$ (see Table 6.1), by artificially growing the declared profit NPA A gets a lower net benefit $\Pi_A = 7 - 3 = 4$. On the other hand, by declaring a lower profit than the true one the risk is to get a portion of the CR that does not cover the effective providers costs. A formal proof that within the Clark tax mechanism, revealing true preferences is the dominant strategy is given in [73].

To summarise, by integrating the Clarke tax mechanism within the COB approach, the guarantee is that the dominant strategy for each individual NPA belonging to a given coalition is to reveal its true profit. The size of the tax imposed to a provider estimates how much its presence along the selected end-to-end route lowers the global coalition profit. The net profit of every provider is then obtained

Agent i	$\Pi_i(r_1)$	$\Pi_i(r_2)$	$\Pi_i(r_3)$	$\Pi'(r_1)$	$\Pi'(r_2)$	$\Pi'(r_3)$	Tax for i
NPA A	7	2	2	9	11	12*	3
NPA B	8	2	4	8	11*	10	3
NPA C	1	9	8	15*	4	6	0
tot Π	16*	13	14	–	–	–	–

Table 6.2: *A malicious agent, e.g., NPA A, by declaring a higher profit in order to influence the end-to-end route selection has to pay a higher Clarke tax. This can reduce the net provider benefit down to a level that does not guarantee to cover the effective costs.*

by subtracting to the revenue the sum of the costs plus the Clarke tax.

In order to implement the Clarke tax mechanism there are however some major issues that need to be considered.

- The *computation of taxes*: all possible alternative situations (i.e., end-to-end routes) must be computed by the CLA in order to compute the taxes for all the NPAs in the coalition. This could become very complex when considering large coalitions of providers with large feasible sets. For this reason, it is essential to eliminate beforehand inconsistent solutions by running the node and the arc consistency steps (see Section 4.3.2).

- *Handling the tax waste.* The Clarke tax itself must be wasted in the sense that it cannot be used to benefit the agents whose voting resulted in being assessed [52], [53]. Ephrati and Rosenschein [73] propose as a possible solution to redistribute the tax for the benefit of agents outside the voting group. In the multi-provider context, this does not really make sense. The main idea within the COB paradigm is therefore to use the taxes to compensate the costs of generating and having coalition leader agents. However, this should be carefully considered since the CLA may as well be a malicious agent.

6.4 Individual Bargaining versus Agents Coalitions

Within the COB approach, the coalition leader agent collects the feasible sets from all the providers along the abstract path selected for the allocation of an incoming service demand. Based on that, the CLA defines the least critical end-to-end route r to be offered to the end user. The global price for r is established by considering the criticalness of the various intra-domain routes selected from each feasible set to form r and by following the ACE pricing policy. This policy is the same that individual NPAs adopt in the NPI framework (see Section 5.2.1).

In the following, the relation between an approach based on individual bargaining agents applying a local Blocking Island (BI) decomposition of their own resources and a paradigm in which a central entity controls a BI-based abstraction

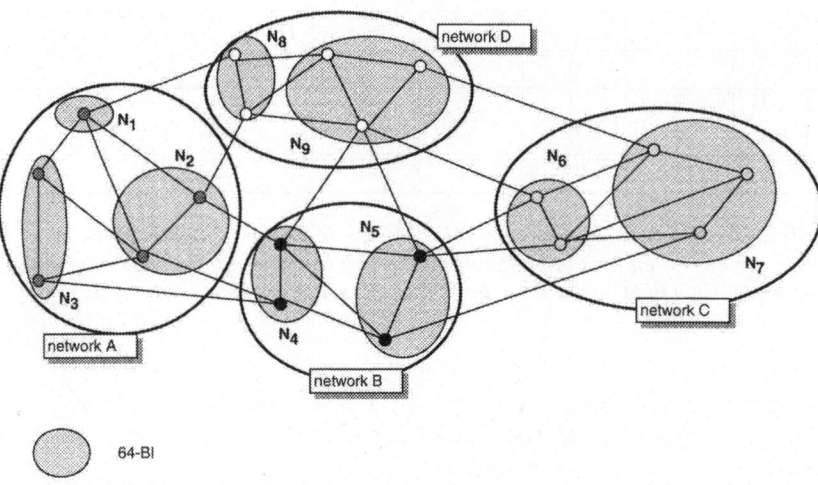

Figure 6.3: *The local BI decomposition of the four distinct networks A, B, C and D makes it possible to cluster nodes and links belonging to the same blocking islands within every domain. In this example, for every provider network the corresponding 64-BIG structure is displayed.*

of the whole multi-provider scenario is investigated. The main objective is to find out if and how the criticalness factor of a specific link varies when passing from a distributed control structure to a centrally supervised scheme. This enables to better estimate and compare the NPI and the COB solutions.

6.4.1 The Local BI Structure versus the Global BI Decomposition

In the multi-provider scenario, distinct operators can make use of blocking islands to represent the bandwidth availability at different levels of abstraction inside their own domains (see Section 3.4.1). A network can be divided up into a set of BIs which cover the whole network graph. This can be used to form an abstract representation of the network that treats β-BIs as single nodes in a graph. Figure 6.3 shows an example where network resources inside four distinct networks A, B, C and D are clustered into the corresponding BIs for the bandwidth requirement $\beta = 64$. The resulting displayed structures correspond to the 64-BIG structures of every different network in the scenario (see Section 3.4.2).

By considering a set of given bandwidth levels $\mathcal{B}^o$, every domain graph $\mathcal{G} = (\mathcal{N}, \mathcal{L})$ can then be represented by the corresponding Differentiated Blocking Island Hierarchy (Diff-BIH) (see Section 3.4.3). This structure can be used for quantitatively estimating the criticalness of network links. The criticalness factor $\mu(l)$ for any link $l \in \mathcal{L}$ in the network graph $\mathcal{G} = (\mathcal{N}, \mathcal{L})$ is formally defined as the inverse of the bandwidth value identifying the father BI of l (see Definition 3.4.8). The lower the father BI of a certain link l is in the Diff-BIH, the less critical is l. In the following, it is investigated if and how the criticalness factor of a specific link varies when passing from a distributed control structure to a centrally supervised scheme. For this purpose, two preliminary concepts are needed.

Definition 6.4.1 (Local BI Decomposition). Given a provider set $\mathcal{P}$, for every distinct network $I \in \mathcal{P}$, the blocking island hierarchy of the graph $\mathcal{G}_I = (\mathcal{N}_I, \mathcal{L}_I)$, for a given set of bandwidth requirements $\mathcal{B}^o$, is called the *local BI decomposition* of the network I.

The local BI decomposition of a certain network I is controlled and managed by the NPA responsible for I. Notice that, when the set $\mathcal{B}^o$ contains only a specific value β, the local BI decomposition of a network I corresponds to its β-BIG.

The group of domains forming a specific provider set $\mathcal{P}$ can be considered as a unique communication network under the control of a central entity, which has a global view of all the resources clustered in it. This global domain can as well be decomposed by using the BI formalism as displayed in Figure 6.4.

Definition 6.4.2 (Global BI Decomposition). Given a provider set $\mathcal{P}$ represented by the graph $\mathcal{G} = (\mathcal{N}, \mathcal{L})$, with $\mathcal{N} = \cup_i \mathcal{N}_i$ and $\mathcal{L} = \cup_i \mathcal{L}_i$ and $\mathcal{G}_i = (\mathcal{N}_i, \mathcal{L}_i)$ the domain graph for every network $i \in \mathcal{P}$, the blocking island hierarchy of $\mathcal{G} = (\mathcal{N}, \mathcal{L})$ for a given set of bandwidth requirements $\mathcal{B}^o$ is called the *global BI decomposition* of the multi-provider network $\mathcal{P}$.

When the set $\mathcal{B}^o$ contains only a specific value β, the global BI decomposition of a $\mathcal{P}$ corresponds to its β-BIG. As it is possible to observe in Figure 6.4, when passing from the local to the global BI decomposition some β-BI may merge. More formally, we have that:

Proposition 6.4.3 (Merging BIs). *Given a network graph $\mathcal{G} = (\mathcal{N}, \mathcal{L})$ and its β-BIG, for any two distinct β-BI_x and β-BI_y, if there exists a link $l \in \mathcal{L}$ interconnecting any node $x \in \beta$-BI_x to a node $y \in \beta$-BI_y with an amount of available bandwidth $\beta(l)$ that is greater than, or equal to, β, then the two distinct blocking islands β-BI_x and β-BI_y must be merged.*

PROOF: This is obvious by applying the route existence, Proposition 3.4 in [93], and the uniqueness, Proposition 3.5 in [93], properties of BIs. $\square$

In order to better understand the consequences of passing from the local to the global decomposition, let us consider the simple example displayed in Figure 6.5. This figure displays the local BI decompositions of the two distinct networks A and B for the bandwidth requirement $\beta = 64$. The network nodes a_3, a_4 and b_1 are boundary nodes, while b_4 is a shared node (see Definitions 3.2.11 and 3.2.12). Merging the two distinct network domains A and B is equivalent to consider a unique network, for instance B, and add nodes and capacity on links that had, before the merging, zero available bandwidth from the perspective of provider B. As Figure 6.6 displays, in the global BI decomposition of the multi-provider network consisting of the domains A and B, there are only two 64-BIs.

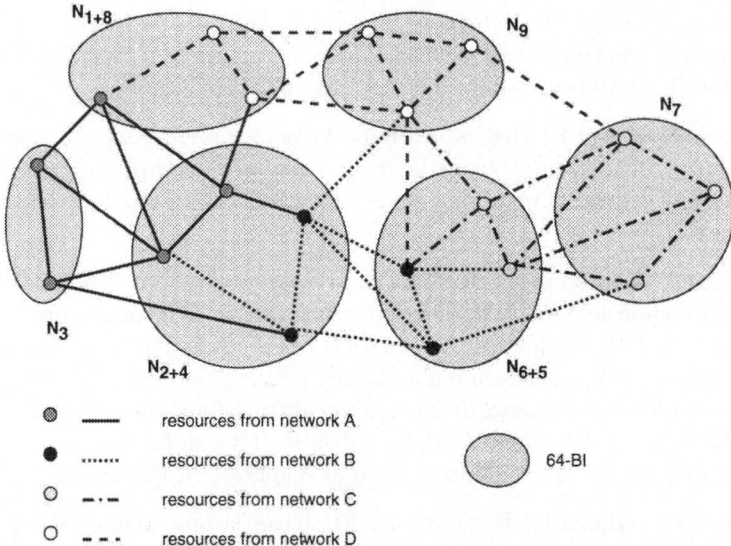

⊙ ▬▬▬	resources from network A	
● ∙∙∙∙∙∙	resources from network B	⬭ 64-BI
○ ▬·▬	resources from network C	
○ ▬ ▬·	resources from network D	

Figure 6.4: *When considering the bandwidth requirement $\beta = 64$, the global blocking island graph, 64-BIG, of the multi-provider scenario displayed in Figure 6.3 consists of six distinct 64-BIs. Three of these BIs have been obtained by merging BIs that were distinct in the local BI decomposition. This happens when at least one of the inter-domain links between distinct BIs, in the local BI decomposition, has an amount of available bandwidth that is greater than, or equal to, the bandwidth requirement $\beta = 64$, see Proposition 6.4.3.*

The main consequence of passing from the local 64-BIG to the global 64-BIG is indeed that from either b_1 or b_2 there exists now a route to either b_4 or b_3 with at least $\beta = 64$ of available bandwidth. This happens because of the increased capacity that, from the perspective of B, is available on the links l_5 and l_6. Because of the route existence and the uniqueness properties of BIs, this means that b_4, b_3, b_1 and b_2 belong to the same 64-BI: N_{2+3+4} (see Proposition 6.4.3).

Therefore, since when merging two networks the main effect of adding capacity is that the network connectivity globally increases:

- A potential higher number of network nodes can be interconnected.

- A higher amount of bandwidth is available on the network links.

6.4.2   The Relativity of Criticalness

Since merging distinct networks globally increase the network connectivity, the criticalness of network resources can change. More precisely, when passing from the local to the global BI decomposition of a certain multi-provider network $\mathcal{G} = (\mathcal{N}, \mathcal{L})$, the criticalness factor $\mu(l)$ of a link $l \in \mathcal{L}$ can either decrease or remain unaltered. This is formally proved by the following theorem.

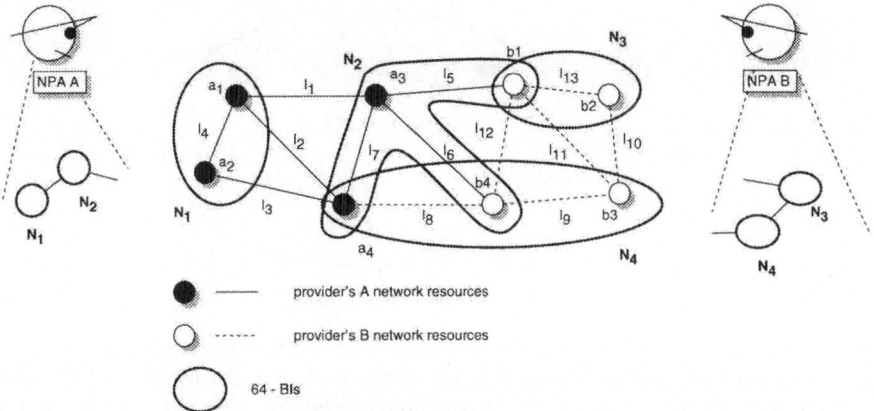

Figure 6.5: *Considering the bandwidth requirement $\beta = 64$, the local BI decomposition of the two networks A and B consists of four 64-BIs. N_1 and N_2 are directly controlled and managed by the NPA A, while N_3 and N_4 are supervised by the NPA B.*

Theorem 6.4.4. *Given a provider set $\mathcal{P}$ represented by the graph $\mathcal{G} = (\mathcal{N}, \mathcal{L})$, with $\mathcal{N} = \cup_i \mathcal{N}_i$ and $\mathcal{L} = \cup_i \mathcal{L}_i$ and $\mathcal{G}_i = (\mathcal{N}_i, \mathcal{L}_i)$ the domain graph for every network $i \in \mathcal{P}$, the criticalness factor $\mu(l)$ of any link $l \in \mathcal{L}_i$ when passing from the local BI decomposition of every network $i \in \mathcal{P}$ to the global BI decomposition of $\mathcal{P}$ cannot increase.*

PROOF: This is obvious by construction of the global BI decomposition of $\mathcal{P}$, when using the same bandwidth requirements considered for building the local BI structures. By adding network nodes and links, the father BI of any link $l \in \mathcal{L}_i$ clustered in a given β_x-BI in the local BI decomposition of any network $i \in \mathcal{P}$ can only become, in the global BI decomposition, a β_y-BI with $\beta_y \geq \beta_x$. Therefore, since the criticalness factor $\mu(l)$ of a any link l is inversely proportional to the amount of bandwidth identifying the father BI of l (see Definition 3.4.8), $\mu(l)$ cannot increase. □

In the example considered in Section 6.4.1, for instance, the respective criticalness factors of links l_1, l_2 and l_3 do not change when passing from the local (Figure 6.5) to the global (Figure 6.6) BI decomposition. On the other hand, for the links l_{10}, l_{11} and l_{12} the criticalness factors decrease. This is due to the fact that in the global BI decomposition these latter links are clustered into a 64-BI, since there exists a route interconnecting their respective end-points with at least 64 available bandwidth. For instance, the route $b_1 - l_5 - a_3 - l_6 - b_4$ guarantees to interconnect the end points of l_{12} with an amount of bandwidth greater than or equal to 82, see Figure 6.6.

When passing from the local to the global BI decomposition, from an individual provider perspective, additional bandwidth can eventually be added only

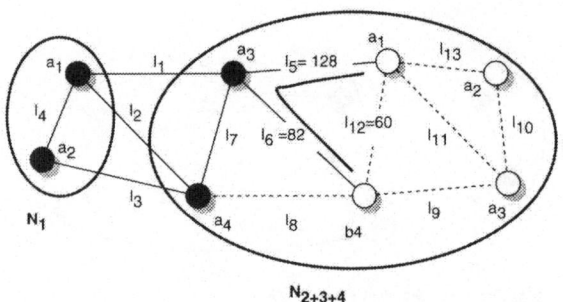

Figure 6.6: *Considering the bandwidth requirement $\beta = 64$, the global BI decomposition of the multi-provider network formed by the domains A and B consists of two distinct 64-BIs. The 64-BI N_{2+3+4} is obtained by merging together the three BIs, N_2, N_3 and N_4, which are distinct in the local BI decomposition as displayed in Figure 6.5. The link l_{12}, for instance, belong to the 64-BI N_{2+3+4}, since there exists a route that interconnects the end points of l_{12} with an amount of bandwidth greater than 64.*

at precise points of the network that are boundary network nodes (see Definitions 3.2.11). Therefore, the effect on the criticalness factor of network links is limited to these areas that are close to the boundary nodes. More precisely, the following theorem formally identifies the links for which the criticalness factor can decrease when passing from the local to the global BI decomposition.

Theorem 6.4.5. *Given a provider set $\mathcal{P}$ and a provider network $A \in \mathcal{P}$ represented by the graph $\mathcal{G}_A = (\mathcal{N}_A, \mathcal{L}_A)$, for any two boundary nodes x and $y \in \mathcal{N}_A$ clustered in two disjoint BIs, $\beta\text{-}BI_x$ and $\beta\text{-}BI_y$, of the local BI decomposition of network A, the links belonging to the intersection of the cocycles of $\beta\text{-}BI_x$ and $\beta\text{-}BI_y$, (if not empty) can become less critical when passing to the global BI decomposition of $\mathcal{P}$.*

PROOF: To simplify the notation $\beta\text{-}BI_x$ is indicated by B_x and $\beta\text{-}BI_y$ by B_y.

Let us assume that an inter-domain link l_x directly connected to the node $x \in B_x$ (or to any other node in the same B_x) and belonging to a distinct provider Y is clustered in a $\beta\text{-}BI_z$ in the local BI decomposition of network Y.

When passing to the global BI decomposition of $\mathcal{P}$, the two blocking islands B_x and $\beta\text{-}BI_z$ (or B_z to simplify the notation) must therefore be merged, see Proposition 6.4.3. Analogously, assuming that an inter-domain link l_y clustered in the same B_z is directly connected to the node $y \in B_y$ (or to any other node in the same B_y), B_y and B_z must as well be merged. Because of the uniqueness property of the BIs, Proposition 3.5 in [93], the three blocking islands B_x and B_y and B_z form a unique B_u in the global BI decomposition of $\mathcal{P}$.

This means that any link $l \in \omega(B_x)$ (or $\in \omega(B_y)$) interconnecting any node $n \in B_x$ (or $\in B_y$) to a node $m \in B_y$ (or $\in B_x$) belong to the B_u, where $\omega(B_x)$ indicates the cocycle of the B_x and $\omega(B_y)$ indicates the cocycle of the B_y. A link $l \in \omega(B_x)$ (or $\in \omega(B_y)$) interconnecting any node $n \in B_x$ (or $\in B_y$) to a node $m \in B_y$ (or $\in B_x$) belong to the $\omega(B_x) \cap \omega(B_y)$, see Definition 3.4.1 of cocycle. Therefore, when passing from the local to the global BI decomposition, the criticalness of any intra-domain link $l \in \omega(B_x) \cap \omega(B_y)$ can only decrease.

□

Summarising, theorems 6.4.4 and 6.4.5 prove that the value of the critical-ness factor of a link can decrease when passing from the local to the global BI decomposition of a network. Hence, for the same multi-provider network scenario and for the allocation of a given demand d, when both adopting the ACE pricing policy as presented in Section 5.2.1, the NPI and the COB approaches may either:

- Define the same global end-to-end route r, in terms of links $l_i \in r$ used, QoS characteristics and global price. This happens when the criticalness of the links $l_i \in r_k$ does not vary when passing from the local to the global BI decomposition.

- Define the same global end-to-end route r, in terms of links allocated $l_i \in r$, QoS characteristics, but with a different global price. This can happen when the criticalness of these links varies when passing from the local to the global BI decomposition. Within the COB approach costs and prices can decrease because of the diminution of the criticalness factor of some links $l_i \in r$.

- Define different global end-to-end routes, which can differ either in terms of the links allocated, QoS characteristics and thereby global price. This can happen when the criticalness of the links selected $l_i \in r$ varies when passing from the local to the global BI decomposition. From the global BI perspective it may be more profitable to allocate links that from the local BI perspective are too expensive.

- Differ for the final outcome of the service set up process. It may happen that while the COB approach succesfully allocates the incoming service demand, the NPI approach fails because self-interested negotiating entities cannot find an agreement despite the existence of solutions able to accommodate the end user request. On the other hand, it may happen that the NPI approach succeeds while the coalition fails. This is due to the fact that when agents dynamically negotiate they do not incur the additional costs and communication overhead, which are due in the COB context to the creation of centrally supervised coalitions.

To conclude, even if the COB approach has from a structural point of view higher chances than the NPI paradigm to succesfully allocate an incoming service demand, the experimental results show that dynamic negotiations are more efficient than coalition based approaches (as demonstrated by the experimental results presented in Section 7.4). The main reasons (further discussed in Chapter 7) can be summarized as follows:

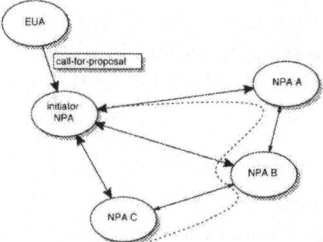

Figure 6.7: *The NPI approach relies upon dynamic bi-lateral interactions between the initiator NPA and the peer providers along the selected abstract path along which the service demand set up process is initiated.*

- Increasing the number of networks does not necessarily increase the solution space.

- There can be a potential increment of conflicts when the average number of interacting self-interested entities increases.

- The adoption of a timeout influences the exploration of the solution space.

6.5   The NPI Paradigm: a Decentralised and Dynamic Approach

The main providers interaction mechanism presented in this volume and discussed in Chapters 4 and 5 relies upon the decentralised and dynamic coordination of distinct self-interested software entities. The end user requesting the allocation of a service demand contacts a specific NPA, the *initiator*, that establishes bilateral agreements with peer providers, see Figure 6.7. In this way, the *initiator* NPA implicitly plays the role of leader within the MuSS process. Even though there is no explicit creation of an organisation, peer providers that decide to collaborate for the allocation of a given service demand implicitly form a coalition. Therefore, the selection of a specific abstract path along which the service demand allocation is started corresponds to the *coalition structure generation* [285]. Within every providers group, individual NPAs are self-interested entities that try to optimise their own profit. The use of of DCSP techniques together with automated negotiations make it possible to find consistent end-to-end routes without the need of revealing strategic information to external competitors. In this way, the repartition of the revenue is automatically done when negotiating peer providers converge to agreements. The prices that they accept to pay to each other guarantee the distribution of the revenue among all of them. In comparison with the FAS and the COB approaches, the main strengths of our decentralised and dynamic architecture can be summarised as follows:

- The definition of a decentralised coordination mechanism reflects the natural distribution of network resources owned and controlled by distinct self-motivated providers.

- The proposed framework addresses the scalability limitations of centralised paradigms. Information describing different domains is locally handled by distinct NPAs reducing the needs of communication.

- Inside its own domain, every provider can perform management, planning, routing and pricing operations independently on the intra-domain mechanisms adopted by peer providers. However, whenever the environment state changes each operator can adapt and modify its control strategies. The environment consists on one side of the network resources under the direct control of a specific operator and, on the other side, of the overall multi-provider market. Software agents are particularly attractive in this context because of their adaptability, i.e., their capability to acquire and process information about the environment they are embedded in.

- The creation of on-demand bi-lateral agreements allows for a more flexible and efficient allocation of resources in individual networks. This has the potential of increasing the provider profit by reducing costs and prices and augmenting the average number of demands succesfully allocated. Static agreements are unable to map the current network state in offers and prices proposed to the end users. Experimental results validate dynamic approaches when compared to solutions based on static agreements (see Section 7.4).

- The amount of information that distinct entities are required to exchange is minimised to boundary constraints[3] and specific offers. In this way, confidentiality of strategic information is preserved and the communication overhead can be reduced.

- If bi-lateral negotiations terminate with an agreement the repartition of the revenue for the service demand allocation is implicitly given by the prices that individual agents accept to pay to each other.

On the other hand, the main critical aspects of such a distributed solution reside in some of the typical drawbacks of decentralised schemes.

- The need of multiple communication flows and service level agreements that are, from the initiator perspective, strictly interrelated introduces interdependencies of different provider-to-provider interactions. This can be very complex to manage unless flexible and dynamic control mechanisms both reactive and pro-active are introduced in every network. When considering this issue, software agents are a very promising technology. Agents perceive their environment and are able to respond in a timely fashion to changes

---

[3]See Section 4.3 for more details on the exchange of boundary constraints.

that may occur in it. In addition, their goal-oriented behaviour offers the capability to anticipate future situations.

- For every peer-to-peer coordination process a certain overhead may slow down the overall service set up process. It is therefore fundamental to ensure an efficient message exchange that is enforced by the use of protocols avoiding loops and deadlocks. For this reason, in the NPI system agents make use of standard and common interaction mechanisms regulated by timeouts.

- Different intra-domain mechanisms and distinct policies may produce globally unstable situations in which it may not be possible to solve conflicts. To overcome this problem, this work proposes the application of DCSP techniques and automated negotiations.

Despite these potential pitfalls, the NPI approach has been validated by experimental results and demonstrated to outperform both the FAS and the COB approaches (see Sections 7.4). A more exhaustive discussion identifying the main strengths and weaknesses of the NPI paradigm, including a comparative analysis of the alternative paradigm discussed in this chapter, is given in Chapter 8.

Chapter 7

Experimental Results

To err is human,
but to really foul things up
requires a computer.
– Murphy's Technology Laws

The NPI approach defines and formalises specific mechanisms for the coordination of multi-domain communication networks. Since these mechanisms are based on a model that simplifies and in some ways idealises a real system (see the main assumptions and the model definition in Chapter 3), their behaviour cannot be completely predicted by a theoretical analysis alone. The principal motivations behind the experimental study presented in this chapter can be summarised as follows:

- There is a large number of interrelated parameters and variables within every provider domain and between distinct networks, and hence a broad range of possible different situations can occur.

- Some parts of the MuSS model that has been defined in this work are heuristic in nature. For instance, since traffic characteristics can only be statistically foreseen or guessed by providers, the activation of specific allocation mechanisms, pricing policies and negotiation strategies cannot be completely anticipated.

- Within the MuSS model, it is possible to influence the end user and provider agents behaviour by specifying needs of consumers and policies of providers at different degrees of detail. Depending on that, the performance of the system can be quite different.

The experimental results reported in this chapter aim to both validate the techniques developed within this context for solving the MuSS problem, and estimate their potential performance and feasibility in realistic frameworks. This comprises two main parts:

1. The specification of the test scenarios, including the definition of specific performance descriptors.

2. The results of two main sets of experiments. The *networking set* estimates typical networking performance. The *microeconomic set* analyses the benefit of the various entities involved in the service provisioning from an economic perspective.

Furthermore, for both the networking and the microeconomic sets, the NPI performances are also compared to the two alternative paradigms implemented for the coordination of network providers: the FAS and the COB methods. The FAS method formalises an automated approach in which coordination is regulated by static long-term agreements (see Section 6.2). The COB paradigm proposes the dynamic coordination of groups of agents by delegating the control of the final phase of the service set up negotiation to a central Coalition Leader Agent (see Section 6.3).

7.1   Experiment Set

The set of experiments that have been run for evaluating the NPI paradigm can be subdivided in two main categories.

- **Networking Set**: These experiments evaluate the feasibility and the efficiency of automated multi-provider coordination mechanisms from a networking perspective. The idea is to estimate the impact of using the mechanisms proposed in this volume in terms of typical parameters used to evaluate network performance [339]. This includes the demand acceptance rate and the average allocation time (see Section 7.2).

- **Microeconomic Set**: This category of experiments analyses the performance of the NPI approach in terms of the end users utility, the providers profit, and the benefit to the society of participants to the service set up process measured by the social welfare. The idea is to evaluate the performance from an economic perspective (see Section 7.3).

In both cases, the results obtained for the NPI approach are also compared to the two alternative procedures implemented for the coordination of network providers: the FAS and the COB approaches.

All the results have been obtained by running the NPI-mas simulator on a local network (10/100 Mb/s-Ethernet). The JATLite *router* run on a Sun UltraSPARC II station with 450 MHz and 512 MB RAM. Agents were randomly distributed in distinct Sun stations with clock rate of 450 MHz and RAM between 256 and 512 MB.

7.1.1 Simulation Set-Up

The multi-provider service set up process involves several entities with different goals and constraints. For the experimental analysis of this complex environment, it has been necessary to make a number of simplifying assumptions (see Chapter 3). Moreover, it has been fundamental to define and identify specific variables that characterise every simulated scenario. This work distinguishes between *independent* and *dependent* experimental variables [222]. Independent variables are the ones whose values are under the control of the experimenter, i.e., they are fixed and known a priori for every experiment. Dependent variables are the variables whose values are not under the control of the experimenter, i.e., their values are observed a posteriori as outcomes of the experiments.

Experimental Independent Components

The assignment of values to independent variables is under the control of the experimenter and represent the input to the simulations. When determing the specific value or the range of possible values that can be assigned to a variable, the experimenter is constrained by limiting the complexity of analysis. In this context, since one of the main objectives of running experiments is to evaluate the feasibility of the NPI approach in realistic environments, a further constraint is to define the variable domains according to realistic frameworks.

Finding real data which describes topologies and traffic profiles in multi-provider environments represented one of the hardest and most unsuccessful tasks of this work. Network and service providers are unlikely to reveal their internal policies and topologies mainly for strategic reasons, which is a common trend given the fierce competition in the current deregulated Telecom market. In addition, for traditional monopolistic (and myopic) reasons many aspects of the resource management, including routing and pricing issues, have not been recorded and documented by many network operators in the past [229]. For this reason, the experimental scenarios and the parameter values chosen for the simulations discussed in this chapter have been drawn by considering examples of existing networks (often simplified and incomplete) found in the Internet or in previous works covering related issues.

For the experiments presented in this chapter, the independent parameters characterising a multi-provider scenario are the following:

- **Network Topology** related parameters. For every simulated provider set $\mathcal{P}$ it is necessary to specify:

    - The inter-domain topology of $\mathcal{P}$ which includes: the number N of distinct provider networks $i \in \mathcal{P}$ and the total number L of inter-domain

Parameter	Description	Possible Values
n	# of nodes per provider network	$\{5, 7, 10\}$
k	# of intra-domain links per provider network	$\{8, 21, 45\}$
N	# of provider networks in $\mathcal{P}$	$\{4, 6, 10\}$
L	# of inter-domain links in $\mathcal{P}$	$\{10, 25, 40\}$
β_c	links capacity (Kbit/s)	$[1000..10000]$
e	end-to-end links delay (μs)	$[2..10]$

Table 7.1: *Network Topology related parameters: For a given provider set $\mathcal{P}$ the corresponding simulation model is described by the distinct network topology parameters summarised in this table (the symbol # stands for "number").*

links. How the distinct networks i are interconnected by these links is randomly defined.

– The intra-domain topology for every network $i \in \mathcal{P}$. This includes the number of network nodes n= $|\mathcal{N}_i|$ and the number of links k= $|\mathcal{L}_i|$, where $\mathcal{G}_i = (\mathcal{N}_i, \mathcal{L}_i)$ is the network graph modelling network i.

For every link, either intra- or inter-domain, the total capacity β_c and the end-to-end delay e characteristics are randomly selected among pre-fixed ranges of possible values (see Table 7.1).

• **Traffic Characteristic** parameters. The simulated traffic profile reproduce on-off Markov processes [291]. Only during 'on' periods end users send service demands. 'On' and 'off' intervals are uniformly distributed with mean of 50 and 10 time units. Moreover, every different simulated traffic profile is characterised by:

– The number δ of service demands per unit of time that are submitted to the provider agents in the scenario (during the 'on' period).

– The total number D of service demands generated per simulation (i.e., over the totality of 'on' intervals).

– The number λ of available service levels and the amount of bandwidth β^o per service level. Given the service model specified in Section 3.2.4, the maximum number of levels considered in the experiments is $\lambda_{max} = 4$, and every bandwidth level β^o expresses a specific value in a given finite and discrete set $\mathcal{B}^o$.

Table 7.2 summaries the different main traffic profiles considered for the tests presented in this chapter. These distinct traffic profiles differ for the number of demands per unit of time and the total number of demands per simulation. For every profile it is then possible to have different distributions of bandwidth requirements based on how many service levels are considered

Traffic Profile	δ	D	λ_{max}	$\mathcal{B}^o$ (Kbit/s)
light	1/15	30	4	$\{19.2, 56, 64, 1000\}$
medium	1/10	50	4	$\{19.2, 56, 64, 1000\}$
strong	1/5	80	4	$\{19.2, 56, 64, 1000\}$

Table 7.2: *There are three main distinct traffic profiles that have been considered in the experiments discussed in this chapter. These profiles have been obtained by varying the number δ of service demands, which are submitted to the provider agents per unit of time, and the total number D of service demands generated per simulation. The maximum number of levels considered in the experiments is $\lambda_{max} = 4$, and every offered bandwidth level is chosen from the set $\mathcal{B}^o$.*

and what is the amount of bandwidth that can be required per level. The *medium* profile has been obtained by doubling both the δ and the total number D of service demands generated with the *light* traffic profile. The *strong* profile has been obtained by doubling both the δ and the D of the *medium* conditions.

- **Negotiation** parameters. This set includes the negotiation duration τ, the selected negotiation protocol (i.e., whether iterations are possible or not) and the type of negotiation strategy adopted by provider agents. The two main NPA strategies that have been tested are the LCF and the MPF procedures (see Section 5.1.3).

- **Pricing Policy** parameters. The pricing policy defines the specific schema, including the type of pricing and cost functions, used by the providers for pricing network resources. In the NPI context, the two main approaches that have been considered are the ACE and the NOC policies (see Section 5.2).

Simulated Network Scenarios

For every simulated provider set $\mathcal{P}$, the possible values of the network topology parameters are constrained by the following simplifying assumptions.

Assumption 7.1.1. There is the same number of nodes n in every provider network $i \in \mathcal{P}$.

Assumption 7.1.2. Every provider network $i \in \mathcal{P}$ is fully connected. This means that for any two nodes in a given network i, there exists an intra-domain link between them belonging to i. Therefore, if n is the total number of nodes in network i, the total number of intra-domain links in i is k $= \frac{n(n-1)}{2}$.

These two first assumptions, guarantee that all providers handle intra-domain search spaces that are equivalent in size. Moreover, in this way, also the maintenance of the BI decomposition for every domain has an equivalent impact on the overall providers performance.

Scenario identifier	n	k	N	L	β_c (Kbit/s)	e μs
S1	5	8	4	10	[1000..10000]	[2..10]
S2	5	8	8	24	[1000..10000]	[2..10]
S3	5	8	12	36	[1000..10000]	[2..10]

Table 7.3: *This table lists the different characteristics of the network scenarios that have been considered for our experiments. For every test-bed the possible values for the links capacity and the end-to-end delay are chosen from the same ranges (see Assumptions 7.1.3 and 7.1.4).*

Assumption 7.1.3. For all simulated scenarios, the bandwidth capacity parameter β_c for both the inter- and the intra-domain links in the scenario is randomly selected among the same discrete and finite set of values (see Table 7.1).

Assumption 7.1.4. For all simulated scenarios, the end-to-end delay parameter e characterising both the inter- and intra-domain links in the scenario is randomly selected among the same discrete and finite set of values (see Table 7.1).

The results reported in this chapter have been obtained for the different simulated network scenarios listed in Table 7.3. The test-beds S1, S2 and S3 mainly differ for the inter-domain topology, i.e., the number N of distinct interconnected provider networks and the number L of inter-domain links. A number of test-beds for the valuation of the relationship between the NPI paradigm and the intra-domain topology have been recently started. However, the results are still premature to be discussed in this volume.

Experimental Dependent Components

The output parameters describing the relevant behaviour and the properties of a given system are indicated as the dependent experimental components. One of the main objectives of running several experiments is to establish which relation may exist between independent and dependent variables. Since the output parameters give a direct feedback of the coordination techniques and of the agents strategies, agents can re-use this feedback in order to dynamically adapt their behaviour. From the providers perspective, for instance, the relevant parameters are the profit, the number of demands succesfully allocated (the better the network resources are allocated the higher is the number of demands that can be accepted) and the average time needed to allocate incoming demands. The fastest a provider can react to demands the more it increases the end user satisfaction and the chances that potential customers will not go toward competitors. For the end users the relevant variables are the utility, the provider response time (or average allocation time) and the provider availability. The availability can be measured, for instance, as the average acceptance rate per specific provider. The more a provider has been available in the past, the more the more it will be preferred over alternative possible operators.

In the NPI context, the set of experimental dependent variables (or simulations output) consists of two main groups of parameters. This subdivision reflects the distinction between experiments focusing on networking issues (the *networking set*) and experiments focusing on remunerative aspects (the *microeconomic set*).

- The **Networking Output Parameters** set includes:

    - The average acceptance rate.
    - The average allocation time.

- The **Microeconomic Output Parameters** set consists of:

    - The end user utility (see Definition 5.1.7).
    - The network provider profit (see Definition 5.1.10)
    - The social welfare (see Definition 5.1.11).

While the microeconomic output parameters have been already defined in Chapter 5 and more details are given in in Section 7.3, the dependent networking variables are formally specified in the following.

Definition 7.1.5 (Average Acceptance Rate). For a given simulated multi-provider network scenario $\mathcal{P}$, the average acceptance rate Ar corresponds to the number of demands successfully allocated over the total number of service demands treated by the providers during the whole simulation period.

$$Ar = \frac{D_s}{D}$$

where D_s indicates the total number of service demands successfully allocated and D the total number of demands received by the totality of providers in $\mathcal{P}$ during the whole simulation period.

As discussed in Section 5.1.4, a service demand is said to be successfully allocated when the negotiation process between the *initiator* NPA and the EUA terminates with an agreement. This can only happen when (1) there exist enough available resources to guarantee the required QoS for the overall service duration in every network along the selected abstract path; and (2) the distinct providers involved in the service set up process converge to common agreements when negotiating with each other. The range of possible values for Ar is $[0..1]$.

Definition 7.1.6 (Average Allocation Time). For a given simulated multi-provider network scenario $\mathcal{P}$, the average allocation time Ta is the average time needed to find an agreement between end users and providers for the allocation of a generic service demand, i.e.,:

$$Ta = \frac{\sum_{i=1}^{D} \theta_i}{D}$$

where θ_i measures the time elapsed between the point in time where a service demand $d_i \in$ D is submitted by an EUA and and the point in time where the *initiator* NPA set up the service, and D is the total number of demands received by the totality of providers in $\mathcal{P}$ during the whole simulation period.

The output data produced by every experiment is a collection of different text files where details about the various processes an agent has been involved in are recorded. For instance, more details about each step of a specific service set up process include timing information, content of the exchanged messages, possible offers, etc. These text files are processed by Perl scripts in order to gather the most significant parameters describing the experiment.

Output Analysis

The absolute values obtained for the experimental dependent parameters should be considered as a quite rough estimation of what could be effectively achieved in real networks. A better evaluation of the NPI performance in real environment would require additional tests with more realistic input data (see Section 7.5).

However, the performance of the techniques considered in this context have been more precisely evaluated in terms of a *confidence interval*, e.g., "we are 95% confident that the expected allocation rate when deploying the NPI paradigm falls in the interval [0.9,1]".

In order to define the notion of confidence interval, some basic notions are briefly recalled in the following (for a more detailed description about systems performance analysis we recommend [147]). When the observed output parameter o is a linear function of a Gaussian process, then o has a normal distribution and the construction of the confidence interval can be established by using the *mean value* $\bar{o}$ of o and the *variance* σ^2 of the N samples. The square root σ of the variance is called standard deviation. Whenever the standard deviation is estimated (like in our case), the Student's t-distribution (rather than the normal one) is used and the form of the confidence interval (CI) is:

$$[\bar{o} - k(\sigma/\sqrt{N}), \bar{o} + k(\sigma/\sqrt{N})] \tag{7.1}$$

with k the appropriate value for the t-distribution[1], $N-1$ degrees of freedom and:

$$\bar{o} = 1/N \sum_{i=1}^{N} o_i \tag{7.2}$$

$$\sigma^2 = \frac{1}{N-1} \sum_{i=1}^{N} (o_i - \bar{o})^2 \tag{7.3}$$

---

[1] The values for k have been computed by making use of the Java script developed by J. Pezzullo and available on-line, http://members.aol.com/johnp71/pdfs.html

The probability associated with the condition $inf < \bar{o} < sup$, where inf and sup identify the CI (as indicated in equation 7.1), is called the confidence level $1 - \alpha_l$. This means that with probability $1 - \alpha_l$ the true value of o will be included in the CI defined by $[inf, sup]$. As observed in [31], the confidence interval is one of the most descriptive measure of the quality of a descriptor (i.e., independent parameter) since it quantifies the measure of spread with an associated probability.

The evaluation of the NPI paradigm has been done by considering the same network scenario (i.e., same independent parameters) for different sets of randomly generated traffic demands. At least 8 distinct tests have been run for every scenario and for every distinct traffic profile (i.e., light, medium or strong). This means that we considered a N included in the range [240..640] depending on the traffic conditions.

7.2 The Networking Results for the NPI Paradigm

The results reported in this section are the outcome of tests performed for the simulated scenarios listed in Table 7.2 when using the NPI paradigm. Within the networking set, the two main output parameters considered to evaluate the NPI performance are the average acceptance rate Ar and the average allocation time Ta.

In the communication network field, automated solutions for the multi-provider service setup are expected to take place over larger time intervals than the ones required for inter-domain routing updates, i.e., $\gg 30$ seconds [78]. In this volume, the implementation and the simulation of alternative coordination approaches such as the NPI, the FAS and the COB procedures enable a more concrete evaluation of relative performance of different automated paradigms.

7.2.1 The Networking Performance for Different Multi-Provider Topologies

In order to estimate the performance of the NPI approach in terms of Ar and Ta against the inter-domain topology, i.e., the number N of provider networks in the scenario and the number L of inter-domain links, the three distinct scenarios S1, S2 and S3 summarised in Table 7.3 have been tested. The scenario S2 has been obtained by doubling the number N1=4 of providers clustered by S1 and passing from L1=10 to L2=24 inter-domain links. The scenario S3 has been obtained by tripling N1 and passing to L3=36 inter domain links.

- *Average Acceptance Rate.* The behaviour of the NPI coordination paradigm when varying the complexity of the inter-domain network topology, is summarised by the results reported in Table 7.4.

    - Traffic *light*. Ar decreases when passing from small multi-provider scenarios, like S1, to networks with a higher number of interconnected

providers like S2 and S3. In comparison with the performance obtained
for S1 (i.e., $Ar = 96\%$), there is in S2 a 2.9% increase in the number of rejected demands. The degradation is higher for the scenario S3,
where there is on the average a 4% increment in the number of rejected
demands.

– Traffic *medium*. Ar decreases when increasing the inter-domain topology (both in the number of domains and links). In comparison with the
performance obtained for S1 (i.e., $Ar = 94.8\%$), 3.3% more demands
are rejected in scenario S2. The degradation is higher for the scenario
S3, where there is on the average a 3.8% increase in the number of
rejected demands.

– Traffic *strong*. As in the previous cases, Ar decreases when increasing
the number of domains in the multi-provider scenario. In comparison
with the performance obtained for S1 (i.e., $Ar = 92.3\%$), there is a
1.2% increment in the average number of demands rejected in scenario
S2 and a 2.3% increase in scenario S3.

- *Average Allocation Time*. The performances of the NPI coordination paradigm in terms of average allocation time Ta, when varying the number N of
interconnected domains, are reported in Table 7.5.

  – Traffic *light*. Ta increases when passing from small multi-provider scenarios, like S1, to networks with a higher number of interconnected
providers, like S2 and S3. In comparison with the performance obtained
for S1 (i.e., $Ta = 20.09$ seconds), in scenario S2 the allocation of a service demand takes 1.29% seconds more on average. The degradation is
higher for the scenario S3, where there is a 6.3% increase in Ta (with
respect to the Ta obtained in the network S1).

  – Traffic *medium*. The allocation of a service demand takes longer on
average in larger multi-provider networks. In comparison with the value
$Ta = 23.52$ seconds obtained for S1, the allocation of a service demand
in S2 takes $Ta = 23.96$ seconds (i.e., 1.87% longer). The degradation is
higher for the scenario S3 where there is a 2.2% increase in Ta (with
respect to the performance obtained in S1).

  – Traffic *strong*. Similarly to the previous cases, the allocation of a service
demand takes longer on average in larger multi-provider network. In
comparison with the value $Ta = 25.70$ seconds obtained for S1, the
allocation of a service demand takes 0.77% more seconds on average in
S2 than in S1. The degradation is higher for S3 where there is on average
a 1.75 % increment in the value of Ta with respect to the performance
in S1.

scen.	Ar (light)			Ar (medium)			Ar (strong)		
	σ	$CI_{99\%}$	mean	σ	$CI_{99\%}$	mean	σ	$CI_{99\%}$	mean
S1	0.05	[0.951,0.968]	0.960	0.04	[0.943,0.954]	0.948	0.06	[0.918,0.928]	0.923
S2	0.08	[0.917,0.944]	0.931	0.09	[0.901,0.926]	0.915	0.07	[0.899,0.911]	0.910
S3	0.06	[0.911,0.929]	0.920	0.10	[0.899,0.920]	0.911	0.08	[0.894,0.905]	0.900

Table 7.4: *Statistics on the acceptance rate for simulations of the scenarios S1, S2 and S3 under different traffic conditions. The confidence level for computing the confidence interval is* $1 - \alpha_l = 99\%$.

scen.	Ta (sec.) (light)			Ta (sec.) (medium)			Ta (sec.) (strong)		
	σ	$CI_{99\%}$	mean	σ	$CI_{99\%}$	mean	σ	$CI_{99\%}$	mean
S1	4.05	[19.9,21.2]	20.09	5.72	[22.80,24.22]	23.52	8.36	[25.12,26.44]	25.70
S2	5.40	[21.1,22.9]	20.35	7.96	[23.27,24.64]	23.96	5.64	[25.12,26.40]	25.90
S3	4.87	[21.0,21.7]	21.37	6.00	[23.26,24.82]	24.04	10	[25.48,27.13]	26.15

Table 7.5: *Statistics on the allocation time for simulations of the scenarios S1, S2 and S3 under different traffic conditions. The confidence level for computing the confidence interval is* $1 - \alpha_l = 99\%$.

Figure 7.1 summarises the average acceptance rate (*case (a)*) and the average allocation time (*case (b)*) against inter-domain topology under different traffic conditions. Both figures are averaged over 20 distinct simulated test runs.

Interpretation ¿From the main experimental results describing the *Ar* and the *Ta* behaviour against the inter-domain topology, it is possible to observe that:

- The average acceptance rate decreases when passing from smaller multi-provider scenarios, like S1, to larger ones, like, for instance, S2 or S3. In larger scenarios, indeed, the average number of providers involved in the allocation of a specific service demand increases, since the end-to-end routes become longer on average. This augments the probability of rejecting an incoming demand for two main reasons. First, a potential higher number of self-interests may come in conflict, second, there is a higher probability of having to cross a domain that has not enough resources to satisfy the demand (i.e., the solution space does not necessarily increases when adding domains). Furthermore, even if alternative paths can be explored it may take too long before finding an agreement among the providers. It would be interesting to run additional experiments in order to find out whether by extending the average negotiation timeout the difference of performance in scenarios of different size would decrease.

- The allocation of a service demand takes longer on average in larger multi-provider networks. This can again be explained by considering that (1) the average number of providers involved in the service set up process increases in

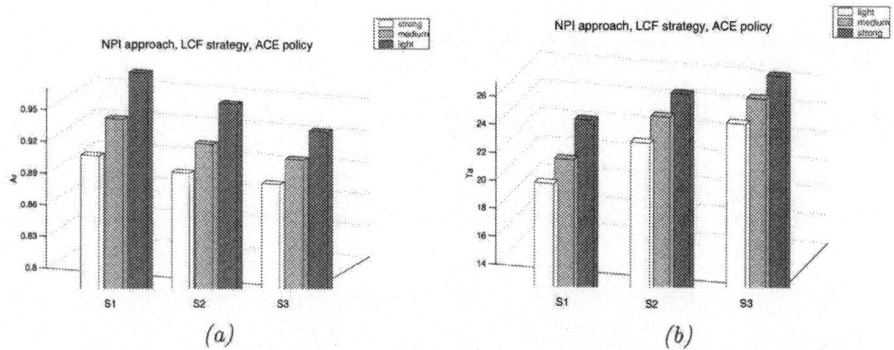

Figure 7.1: *Average acceptance rate (a) and average allocation time (b) obtained for multi-provider networks of different size. When passing from S1 to S2 the number N of distinct networks double and when passing from S1 to S3 N triple. The three scenarios have been tested under different traffic profiles when provider agents make use of the LCF strategy and of the ACE pricing policy.*

more populated domains and (2) the search space for every distinct operator increases (as a consequence of a higher number of inter-domain links). Hence, reaching an agreement satisfying all providers involved in the service set up may require more time.

- The difference in terms of performance among the multi-provider networks of different size decreases for heavier traffic loads. This means that the impact of higher traffic volumes is stronger for scarcely populated domains. This is due to the fact that, on the average, the increment in the number of demands δ per unit of time that every provider has to process is lower in larger domains.

- The scalability of the NPI paradigm can be better quantified by considering that:

  - When doubling the multi-provider network size (i.e., when passing from S1 to S2), there is a 2.9% maximum degradation of Ar (that happens for light traffic conditions) and a 1.29% maximum degradation in terms of average allocation time (that happens for medium traffic load).

  - When tripling the dimension of the scenario (i.e., when passing from S1 to S3), there is a 4% maximum degradation of Ar (that happens for light traffic conditions) and a 6.3% maximum degradation in terms of average allocation time (that happens for light traffic conditions). Notice that in this latter case, the difference is 1.28 seconds.

7.2.2 The Networking Performance for Different Traffic Profiles

This section focuses on the results obtained for the NPI paradigm with providers making use of the LCF strategy and of the ACE pricing policy when traffic conditions vary in the network. The main idea is to investigate how the developed coordination techniques behave when the average number of allocation requests per unit of time received by a network provider increases.

The average acceptance rate Ar and the average allocation time Ta performance have been investigated for the three distinct scenarios S1, S2 and S3. Each distinct multi-provider network has been tested under three main different traffic conditions described by the profiles summarised in Table 7.2. The results reported in the following are summarised in Tables 7.4 and 7.5.

Scenario S1

- The average acceptance rate Ar decreases when increasing the intensity of the traffic in the multi-provider network.

  - On the average the 96% of the service demands is successfully allocated for the *light* traffic profile.

  - When doubling the intensity (*medium* profile), the value of Ar decreases to an average of 94.8%. There is a 1.2% decrement, in comparison with the performance obtained for the *light profile*.

  - For *strong* traffic conditions, the percentage of demands successfully allocated is 92.3%. In comparison with the performance obtained for the *light* profile, 3.7% more demands are rejected on average.

- The average allocation time Ta increases when increasing the intensity of the traffic injected in the multi-provider network.

  - For the *light* traffic profile the average Ta is equal to 20.09 seconds.

  - When doubling the traffic intensity, the value of Ta increases up to 23.52 seconds. There is a 17.07% increment in Ta, in comparison with the performance obtained for the *light* profile, assumed as the reference situation.

  - For *strong* traffic conditions, Ta increases on the average up to 25.70 seconds. In this case, there is a 27.9% increment in Ta when comparing to the performance obtained for the *light* traffic load.

Scenario S2

- The average acceptance rate Ar decreases when increasing the intensity of the traffic load in the scenario.

- On the average, the 93.1% of the service demands is successfully allocated for *light* traffic conditions.

- When doubling the intensity (*medium* profile), the value of Ar decreases to an average of 91.5%. This means that 1.6% more demands are rejected in comparison with the *light* case.

- In the case of *strong* traffic conditions, the percentage of demands successfully allocated is 91%. The decrement in Ar is 0.5% when comparing the performance obtained for the *light* profile.

• The average allocation time Ta increases when increasing the intensity of the traffic.

- For the *light* traffic profile the average Ta is equal to 20.35 seconds.

- When doubling the traffic intensity, the value of Ta increases up to 23.96 seconds. There is a 17.73 % increment in Ta, in comparison with the performance obtained for the *light* traffic load.

- For *strong* traffic conditions, Ta is on average 25.9 seconds and, in comparison with the *light* profile performance, it takes 27.2% longer to allocate a demand.

Scenario S3

• The average acceptance rate Ar decreases when increasing the intensity of the traffic in the scenario.

- On the average, the 92% of the service demands are successfully allocated for the *light* traffic profile.

- When doubling the intensity (*medium* profile), the value of Ar decreases to an average of 91.1% of demands succesfully allocated. The decrement in Ar, in comparison with the performance obtained for the *light* traffic conditions, is 0.9%.

- Passing to the *strong* traffic conditions, the percentage of demands successfully allocated is 90%. On the average, 1% more demands than for the *light* load are rejected.

• The average allocation time Ta increases when increasing the intensity of the traffic.

- For the *light* traffic profile the average Ta is equal to 21.37 seconds.

- When doubling the traffic intensity, the value of Ta increases up to 24.04 seconds. In comparison with the *light* performance, it takes 12.49% longer on average to allocate a demand.

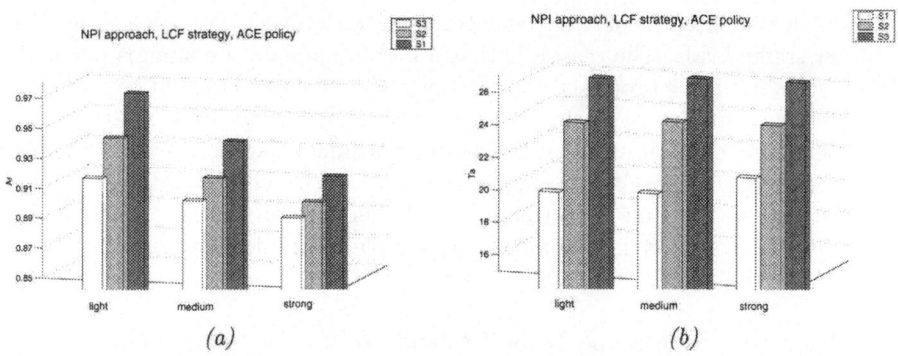

Figure 7.2: *Average acceptance rate (a) and average allocation time (b) obtained for the three scenarios S1, S2 and S3 when running three different traffic profiles and making use of the NPI approach for providers making use of the LCF strategy and of the ACE policy. The average number of service demand requests generated per unit of time doubles when passing from light to medium traffic conditions and triples when passing from the light to the strong profile.*

- For *strong* traffic conditions, Ta is 26.15 seconds and the increment in Ta, in comparison with the performance obtained for the *light* load, is 22.64% in this case.

Figure 7.2 summarises the average acceptance rate (*case (a)*) and the average allocation time (*case (b)*) against the traffic conditions for the three distinct scenarios S1, S2 and S3. The figures are averaged over 20 distinct simulated test runs.

Interpretation The main experimental results describing the Ar and the Ta behaviour against various traffic conditions can be summarised and interpreted as follows:

- In all simulated scenarios, the average acceptance rate Ar decreases when traffic conditions become heavier. This is mainly due to the incapability for network providers to satisfy an increasing traffic volume, given that the amount of available inter- and intra-domain resources is finite.

- The degradation in terms of Ar for higher traffic load is lower when passing from multi-provider scenarios scarcely populated, like S1, to networks with a higher number of interconnected providers like S3. This is due to the fact that in larger scenarios the increase of traffic is distributed over a larger number of distant providers. Therefore, the number of demands per unit of time δ that each provider has to process is on the average lower.

- In all simulated scenarios, the average allocation time Ta increases for heavier traffic loads. The reason is that a higher number of demands per unit of time δ has to be processed by every provider. This effect could be reduced by having a specific agent for each distinct demand. However, multiple NPAs per single domain would still have to coordinate themselves when requiring and reserving specific resources in the network. This would have a direct impact on the time performances and therefore on the acceptance rate as well. This aspects in discussed as a possible future development of this work (see Section 8.3).

- The degradation of the Ta performance for heavier traffic conditions is decreasing when passing from scenarios scarcely populated to networks with a higher number of interconnected providers. The main reason is again that on average the increment on the number of generated demands δ per unit of time, for heavier traffic conditions, is lower for larger networks.

- Despite the lack of more realistic input data concerning both traffic and network topologies for a precise estimation of the performance in real environments, the fundamental result is that:

  – When passing from the *light* to the *medium* traffic profile, for every simulated scenario, the degradation in term of Ar falls in the range that goes from 0.9% to 1.6%. When passing to the *strong* traffic conditions, the decrement in Ar falls in the range between 1.38% and 4.2%. This means that, over the three simulated scenarios, when tripling the traffic load, there is still an average of 90.5% of demands that can still be successfully allocated.

  – The impact of heavier traffic load on Ta is also relatively reduced. When doubling the traffic load there is a average increase included in the range [11%,17.54%] in the time required for allocating a demand. This is due to the larger number of parallel process that distinct providers are involved in. When tripling the traffic load there is a average increase in Ta falling between 19.38% and 26.53%. In absolute terms, the differences are in the order of seconds (the maximum difference observed is in S1 when tripling the traffic load is for instance 5.3 seconds).

7.2.3   The Impact of Negotiation Strategies on Networking Performance

As discussed in Chapter 5, both end user and provider agents follow specific strategies when negotiating for the allocation of service demands. When varying these strategies, negotiation participants influence the outcome of the negotiation process.

The risk attitude of an EUA is summarised by the way the service is evaluated. As discussed in Section 5.1.4, given the valuation function $V(d,t) = p_r^u(d) \times \sigma(t)$ for a demand d at time t, different users attitudes are specified by different shapes of the curve $\sigma(t)$. In the experiments presented in this chapter, the following generic service valuation function is considered:

$$V(d,t) = p_r^u(d) \times (t/\theta_{neg})^\alpha$$

where $p_r^u(d)$ is the reserve price of the end user for demand d, θ_{neg} is the negotiation timeout and $\alpha = 1$ for risk neutral EUAs, $\alpha < 1$ for risk averse EUAs and $\alpha > 1$ for risk seeking EUAs. In all the tests performed, the risk attitude of the EUAs populating every scenario has been randomly determined. The fundamental idea is that a random distribution of the risk attitude of potential customers more closely correspond to real heterogeneous markets[2].

The two main possible NPA strategies that have been considered in the experiments are the MPF and the LCF strategies (described in Section 5.1.3). In the latter case, among its feasible offers, a NPA selects the one that maximises its profit. Then, if more than one offer produce the same benefit, the NPA selects the option requiring the least critical resources. With the LCF strategy, among the feasible offers, a NPA gives the priority to the least critical choice. Then, if more alternatives are characterised by the same degree of criticalness, the NPA chooses the most profitable option.

All results reported here have been obtained by averaging over 8 distinct test runs differing for the set of randomly generated demands D_i, with $i \in \{1, \ldots, 8\}$. Each set of demands has been generated under *medium* traffic conditions. For each set D_i the same network scenario has been populated once with agents adopting the LCF strategy and then with provider agents using the MPF strategy. Figure 7.3 displays the results obtained for the scenario S1.

Scenario S1

- *Average Acceptance Rate.*

  When providers are using the LCF strategy the average allocation rate is $Ar = 94.20\%$. When adopting the MPF strategy the Ar decreases down to 92.78%. There is 1.42% average difference in Ar between the two strategies.

- *Average Allocation Time.*

  On the average, the allocation time is higher when the provider agents adopt the MPF strategy: Ta is equal to 22.93 seconds when making use of the LCF strategy and to 24.06 seconds when adopting the MPF strategy. There is a 4.92% difference (i.e., 1.13 sec) in Ta between the two strategies on average.

---

[2]Notice that the usage of different end user strategies and risk attitudes is discussed in Section 9.3 as a possible future extension of the work presented in this volume.

Scenario S2

- *Average Acceptance Rate.*

  The average acceptance rate when provider agents make use of the LCF strategy is $Ar = 90.82\%$. When adopting the MPF strategy the Ar decreases down to the 87.17%. On the average, there are 3.65% more demands that can be allocated when adopting the LCF strategy.

- *Average Allocation Time.*

  The average allocation time is higher when the provider agents make use of the MPF strategy: Ta is equal to 24.14 seconds for the LCF strategy and to 25.30 seconds when adopting the MPF strategy. On the average, there is a 4.80% difference in Ta (i.e., 1.16 sec) between the two strategies.

Scenario S3

- *Average Acceptance Rate.*

  When providers are using the LCF strategy the average allocation rate is $Ar = 93.38\%$. When applying the MPF strategy the Ar decreases down to the 88.49%. On the average, there is 4.89% difference in Ar between the two NPA strategies.

- *Average Allocation Time.*

  On the average, the allocation time becomes higher when the provider agents make use of the LCF strategy: Ta is 27.10 seconds for LCF and 28.35 seconds for the MPF strategy. There is 4.61% difference in Ta (i.e., 1.25 sec) on average between the two approaches.

Interpretation   Gathering the experimental results for the three different scenarios reported above and considering more in details the simulation output (e.g., the failure reasons messages exchanged by agents) it is possible to give the following explanation of the observed performance.

- The use of the LCF strategy outperforms the adoption of the MPF method in terms of average acceptance rate Ar. Concerning the average allocation time, in larger scenarios like S3, the MPF option becomes faster than the LCF approach. However, the reason is that while with the MPF strategy a higher number of processes terminate earlier because of the incapability of satisfying higher prices, the LCF strategy makes it possible to terminate more allocation processes (lasting longer on average than the failed set ups). These results can be better explained considering the following factors.

  - Since the profit is directly proportional to the price asked for the allocation, the first offered prices are higher on average when adopting the MPF strategy. Therefore, the probability that a higher number of proposals may be rejected by the end users or peer providers increases. This is confirmed by the agents failure reasons observed.

- Since prices are higher on average with the MPF strategy, the negotiation between the initiator NPA and the EUA requesting the set up of a certain service demand could go through several iterations before the agents find a mutual agreement. Thus, the chance that the negotiation timeout is reached and the service allocation fails increments. This is also explaining why the average allocation time is higher with the MPF approach (at least in scenarios S1 and S2).

- When passing from scenario S1 to larger networks like S2 and S3, the average allocation time increases more in the case provider agents apply the LCF approach. This is mainly due to the fact that there is a higher number of service demands on average succesfully allocated than with the MPF strategy. These successful allocations last longer than processes that do not terminate because of some failure: when adopting the MPF approach a higher number of processes on average terminate earlier because it is not possible to find an agreement between peer providers and end user.

• The difference of performance in terms of *Ar* increases when passing from S1 to larger networks like S2 and S3. By examining the output simulation files and the increasing average degradation in the number of demands succesfully allocated when passing to larger scenarios (see Section 7.2.1), it is possible to give the following explanation. In larger networks, a potential higher number of self-interested entities may have to cooperate for the allocation of a specific demand. In a context where providers offer on the average higher prices (MPF approach), it is even more difficult (than in the LCF case) to find an agreement satisfying all the negotiating entities. The existence of more alternative paths (in larger scenarios) does not seem to compensate the higher number of negotiation failures.

More exhaustive conclusions are drawn in Section 7.3.2, after having analysed the impact of the two different strategies on the utility, profit and social welfare.

7.2.4 The Impact of Pricing Policies on the Networking Performance

The way costs and prices are estimated in every network has an impact on the control in the network of resources utilisation and therefore on how the NPAs make offers to potential customers (see Section 5.2.1). In the NPI framework, two main distinct mechanisms for assigning costs have been considered.

• The ACE policy makes use of a cost function that dynamically evaluates the opportunity cost of allocating specific links, taking into account their

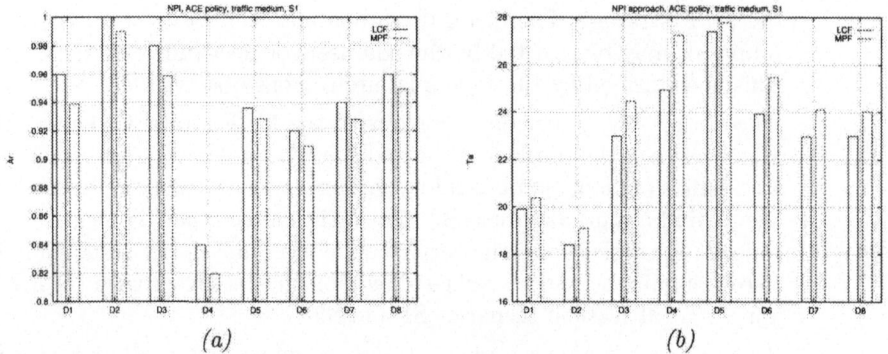

Figure 7.3: *Average acceptance rate (a) and average allocation time (b) obtained for the scenario S1 under medium traffic conditions for providers making use of the LCF and MPF strategies.*

criticalness (see Definition 5.2.12). At the same time, also resource prices are correlated to the criticalness factor of the links forming the selected intra-domain route (see Equation 5.3). Criticalness is in this case related to the BI decomposition of network resources. In this way, the profit, computed as the difference between prices and costs, is directly proportional to the criticalness of the resources allocated.

- With the NOC policy, costs are assumed to be constant values that every provider internally establishes. Moreover, resource prices are computed by considering the amount of available bandwidth $\beta(l)$ on any link l allocated (see Equation 5.4). With this latter approach, criticalness is not related to the BI decomposition of the network, but simply to the amount of capacity available on the links.

All results reported in the following have been obtained by averaging over 8 distinct test runs differing for the set of randomly generated demands D_i, with $i \in \{1, \ldots, 8\}$. Each set of demands has been generated under *medium* traffic conditions. For each set D_i, the same network scenario has been populated once with agents making use of the ACE policy and once with agents adopting the NOC policy. Figure 7.4 displays the results obtained for the scenario S1.

Scenario S1

- *Average Acceptance Rate.* When providers are using the ACE policy the average acceptance rate is $Ar = 94.20\%$. When relying upon the NOC policy the Ar decreases down to the 82.47% of demands succesfully allocated. There is a 11.73% average difference in Ar between the two strategies.

- *Average Allocation Time.* On the average, the allocation time is higher when the provider agents make use of the ACE policy: Ta decreases from an average of 22.93 to 21.01 seconds when adopting the NOC policy. There is a difference of 1.92 seconds on average (i.e., 9% increase in Ta when passing from the NOC to the ACE policy).

Scenario S2

- *Average Acceptance Rate.* When provider agents adopt the ACE pricing policy, the average number of demands succesfully allocated is $Ar = 90.82\%$. When applying the NOC policy the Ar decreases to the 78.5%. There is therefore a 12.32% difference on average between the two approaches.

- *Average Allocation Time.* The average allocation time is still higher when making use of the ACE policy. For the NOC pricing approach the observed value of Ta is 23.03 seconds, while when dynamically computing costs and prices based on the network criticalness the average allocation time is $Ta = 24.14$ seconds. There is a difference of 1.11 seconds on average (i.e., 4.8% increase in Ta when passing from the NOC to the ACE policy).

Scenario S3

- *Average Allocation Rate.* When providers are using the ACE policy the average acceptance rate is $Ar = 93.38\%$. When using the MPF strategy the Ar decreases to 75.53%. On the average, there is therefore a 17.85% difference between the two approaches.

- *Average Allocation Time.* The average allocation time becomes higher when adopting the NOC policy. In this latter case, the observed value of Ta is 28.23 seconds, while when dynamically computing costs and prices based on the network criticalness the average allocation time is $Ta = 27.10$ seconds. There is a 4.16% average difference, this time in favour of the ACE option.

Interpretation Gathering the results for the three distinct experimental scenarios S1, S2 and S3, and examining more in details the messages exchanges by agents, it is possible to explain the different performance of the two distinct policies as follows:

- The ACE pricing policy outperforms the NOC policy in terms of the average number of demands succesfully allocated. This is directly related to the capability of the BI-based criticalness valuation of network resources, used with the ACE policy, of dynamically estimating not only prices but also opportunity costs. This allows a better utilisation of network resources since the ACE costs dynamically reflect the probability of blocking future incoming demands when allocating a certain amount of network capacity, i.e., bandwidth on a link. The criticalness factor (see Definition 3.4.8) used for the opportunity cost computation[3] estimates indeed the risk of isolating

---

[3]See Section 5.2.1 for more details on price and cost computation.

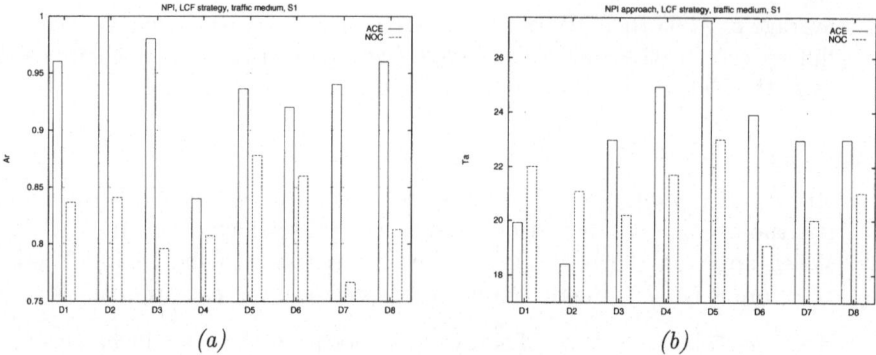

Figure 7.4: *Average acceptance rate (a) and average allocation time (b) obtained for the S1 under medium traffic conditions for providers making use of different pricing policies.*

some network nodes in terms of reachability when allocating the resources on a particular link.

- The gain in terms of Ar when applying the ACE policy increases for scenarios with an increasing number of interconnected networks. This is mainly due to a higher degradation of performance on average when passing from smaller to larger scenarios for providers using the NOC policy. The longer the average paths along which a demand may have to be allocated become, the higher is the negative impact of not computing costs and prices by dynamically taking into account the resource criticalness. As a matter of fact, a higher portion of providers in the scenario does not dynamically estimate the opportunity costs in a way that the probability of blocking future incoming demands can be reduced. This implies a higher rejection rate on average for providers using the NOC policy.

- When provider agents adopt the NOC policy a service demand allocation is faster on average for medium-small scenarios. This is a direct consequence of how costs and prices are estimated within the two different pricing approaches. While within the NOC paradigm, costs valuation is static, within the ACE policy providers dynamically recover information from the Diff-BIH decomposition to dynamically compute the criticalness factor that intervenes in the cost computation. However, in larger scenarios, like S3, the situation is reversed and despite the dynamic computation of costs and prices allocations are faster on average when agents adopt the ACE policy. When considering more in details the simulation output files registered by agents, this seems to be determined by an average higher duration of the negotiation processes when making use of the NOC policy. This is due to the fact that since costs are kept constant, in order to achieve a certain level of profit providers may not be able to optimise the prices. Higher prices may need to be iteratively negotiated with the end users determining a higher duration on average of the allocation process. This is amplified in larger scenarios where the average number of providers involved in the set up process increases.

7.3   The Microeconomic Set

The microeconomic set groups the main results obtained to investigate the performance of the techniques presented in this book from an economic perspective. For this purpose, the experimental dependent parameters observed in this case are the utility, the profit and the social welfare.

For a given service demand d, each EUA computes its utility(see the formal Definition 5.1.7) as the difference between the valuation or reserve price $p_r^u(d)$ of d and the price paid to the NPA to obtain the allocation of d:

$$\mathcal{U}(d) = p_r^u(d) - p(d) \tag{7.4}$$

In all the experiments discussed in the following, for each generated demand there is an individual EUA[4]. For convenience, the values observed for the utility are kept in the interval $[0, 1]$ by dividing every term of equation 7.4 by the valuation of d. This gives the *normal utility* $\mathcal{U}_n$:

$$\mathcal{U}_n(d) = 1 - p(d)/p_r^u(d) \tag{7.5}$$

The boundary $\mathcal{U}_n = 1$ is reached, in principle, only in the hypothetical situation of obtaining the allocation of a service for free. However, this never happens in the NPI framework. On the other hand, because of the assumption that an end user agent accepts a deal only if this does not decrease its own profit, the inferior boundary for $\mathcal{U}_n$ is zero. Since the absolute service valuation, and therefore the absolute value of the utility, are somewhat arbitrary to individual EUAs, the main focus is in the following on relative performance of average values, i.e., differences of utility. The *average* normal utility U per simulated scenario per EUA is computed by dividing the sum of the normal utilities of all users in the scenario (adding zero for every rejected demand) by the total number of generated demands during the experiment (that is equivalent to the global EUAs population).

$$U = \sum_{i=1}^{N} \mathcal{U}_n(i)/N \tag{7.6}$$

where N is the total number of EUAs populating the simulated scenario. Analogously, given that the providers profit (see Section 5.1.3 for the formal definition) $\pi(d)$ is defined as the difference between the revenue $p(d)$ and the total cost $c(d)$ of the resources needed to allocate a given demand d, the *normal profit* π_n is computed as follows:

$$\pi_n(d) = 1 - c(d)/p(d). \tag{7.7}$$

---

[4]In Section 8.2.2, this choice is discussed in more details.

The boundary $\pi_n = 1$ is reached in the hypothetical situation in which allocating a demand has zero cost for all the providers involved in the set up process. However, this is never the case in the NPI framework (see Assumption 5.1.5). On the other hand, because of the assumption that a provider agent accepts a deal only if this does not decrease its own profit, the inferior boundary for π_n is zero. Also for the profit, the main focus is on relative performance of average profit values. The average normal profit π_n^a per NPA is computed by dividing the total profit of the NPA over the whole simulated period divided by the total number of demands received by the NPA (adding zero for every allocation failure). Then, the average normal profit Π per simulated scenario per NPA is obtained as it follows:

$$\Pi = \sum_{i=1}^{N} \pi_n^a(i)/N \tag{7.8}$$

where N is the total number of NPAs populating the simulated scenario.

The normal social welfare $\mathcal{W}_n$ per transaction is given by:

$$\mathcal{W}_n = \sum_{i=1}^{P} \pi_n(i) + \sum_{j=1}^{M} \mathcal{U}_n(j)$$

where P is the total number of providers i involved in the allocation and M is the total number of users j involved in the transaction. The normal social welfare W per simulation is given by

$$W = \sum_{i=1}^{D} \mathcal{W}_n(i)$$

/D where D is the total number of transactions i taking place during the whole experiment. If no deal has been made between the all agents populating the simulated scenario during the whole simulation period, then $W = 0$. The hypothetical upper bound limit $W = 2$ would be reached only when all deals during the whole simulation period are concluded with $U = 1$ and $\Pi = 1$. However, for a balanced satisfaction of both users and providers the average values for the observed W should fall around 1.

7.3.1   Utility, Profit and Social Welfare for Different Traffic Conditions

As discussed in Section 7.2.2, the behaviour of the NPI coordination techniques has been tested under different traffic conditions. This with the aim of observing how a higher average number of allocation requests received by provider agents can impact the performance of the developed techniques. While the effect on the average allocation time and on average number of demands succesfully allocated has been discussed in Section 7.2.2, in this section, we consider the impact on utility, profit and social welfare.

traffic	S1			S2			S3		
	U	Π	W	U	Π	W	U	Π	W
light	0.6258	0.768	1.330	0.6334	0.6625	1.285	0.5104	0.5401	1.048
medium	0.6185	0.756	1.303	0.6100	0.6589	1.270	0.5031	0.5240	1.027
strong	0.6097	0.739	1.289	0.5971	0.6496	1.264	0.4913	0.5080	1.018

Table 7.6: *Statistics on average normal utility, profit and social welfare for simulations of the scenarios S1, S2 and S3 under different traffic conditions.*

The experimental results obtained for the scenarios S1, S2 and S3 and for different traffic profiles are summarised in Table 7.6. This output has been averaged over 20 distinct simulated test runs for every distinct multi-provider configuration.

- *Utility.* For all the three simulated scenarios it is possible to observe that when increasing the traffic load the average normal utility decreases. This is directly related to the fact that when providers receive a higher number of allocation requests per unit of time there is a decrement on average acceptance rate (see Section 7.2.2). Therefore, the average end users satisfaction decreases since the number of users for which a demand cannot be satisfied increases. In fact, in the average normal utility U computation a zero is added for every allocation failure.

- *Profit.* For all the three simulated scenarios, it is possible to observe that when increasing the traffic load the average normal profit decreases. Also this phenomenon is dependent on average decrease of Ar. From the providers perspective, this is mainly due to following factors:

  - Since network resources are finite, when the average number of service demand allocations increase, the bandwidth available for future allocations decrease (i.e., criticalness increase). This increases the opportunity cost of network resources degrading the average providers profits.

  - The amount of computational resources of provider agents is also finite. Therefore, when the average duration of a specific service allocation increases the probability that additional deals can be processed on time decreases.

- *Social Welfare.* As a direct consequence of the average decrease of both the average normal utility and the average normal profit also the average normal social welfare decreases for heavier traffic conditions. This degradation is a general phenomenon that concerns network performance when traffic conditions become heavier [262]. A more precise estimation of the influence on this degradation of the NPI paradigm is possible when considering alternative automated software based approaches like the FAS and the COB paradigms (see Sections 7.4).

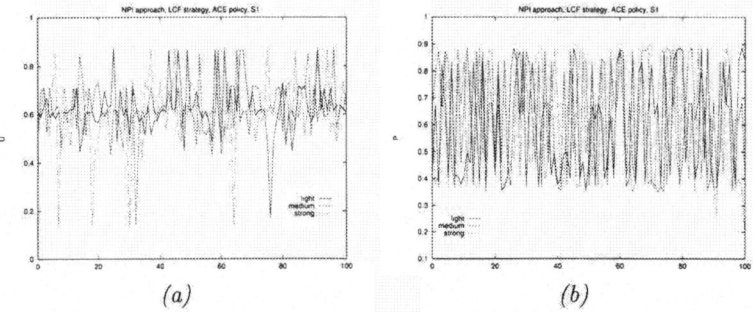

(a)   (b)

Figure 7.5: *Average normal utility (a) and average normal profit (b) against different traffic conditions when applying the NPI approach for providers making use of the LCF strategy and the ACE policy in scenario S1.*

The decrement observed for the average normal utility when passing from S1 to S2 and S3 is due to the fact that on average in larger domains there are higher prices. The reason is that in scenarios with a higher number of interconnected providers a service demands can be allocated over a longer end-to-end route (i.e., involving more resources and more self-interested entities to be satisfied). From the providers perspective the average normal profit decreases in larger domains mostly because a higher number of demands is on average rejected (see Section 7.2.2). These two aspects together explain why also the social welfare decline when passing from S1 to larger multi-provider environments.

Figure 7.5 shows the end users normal utility (*case (a)*) and the normal providers profit (*case (b)*) against different traffic conditions for the three distinct scenarios S1, S2 and S3. These figures are averaged over 20 distinct simulated test runs for a set of 100 randomly generated service demands.

7.3.2 The Impact of Negotiation Strategies on Utility, Profit and Welfare

The networking performances obtained in the NPI context by adopting either the MPF or the LCF strategy are considered in Section 7.2.3. In the following, we illustrate how the use of different strategies influence the outcome in terms of average normal utility, profit and social welfare.

All results reported here have been obtained by averaging over 8 distinct test runs differing for the set of randomly generated demands D_i, with $i \in \{1, \ldots, 8\}$. Each set of demands has been generated under *medium* traffic conditions. For each set D_i the same network scenario has been populated once with agents using the LCF strategy and then with provider agents adopting the MPF strategy. Figures 7.6 and 7.7 display the results obtained for the scenario S1.

Scenario S1

- U: The average normal utility is higher for all the different set of demands when the provider agents make use of the LCF strategy. There is a 9.57% decrease in U when the providers propose first the most profitable offer (i.e., MPF strategy).

- Π: The average normal profit is higher when providers offer first the least critical resources (i.e., LCF strategy). LCF outperforms MPF by a difference of 1.2% on average. For the set of demands D4 and D8, the average normal profit is higher when providers adopt the MPF strategy.

- W: As a direct consequence of the previous results, the average normal welfare is higher when NPI agents make use of the LCF strategy. There is a 6.77% difference in the average normal welfare. For the set of demands D4, the average normal welfare is higher when adopting the MPF strategy.

Scenario S2

- U: The average normal utility is higher for all the different set of demands when the provider agents make use of the LCF strategy. There is a 2.54% decrease in U when the providers submit first the most profitable offers (i.e., MPF strategy).

- Π: The average normal profit is, for all the observed simulations, higher when providers offer first the least critical resources (i.e., LCF strategy). LCE outperforms MPF by a difference of 3.45% on average.

- W: The average normal welfare is higher when provider agents apply the LCF strategy. There is a 9.02% difference in the average normal welfare.

Scenario S3

- U: The average normal utility is higher for all the different set of demands when the provider agents make use of the LCF strategy. There is a 5.03% decrement in U when the providers propose first the most profitable offer (i.e., MPF strategy) instead of the lest critical one (i.e., LCF strategy).

- Π: The average normal profit is, for all the observed simulations, higher when providers offer first the least critical resources (i.e., LCF strategy). LCE outperforms MPF by a difference of 4.67% on average.

- W: As a direct consequence of the previous results, the average normal welfare is higher when NPI agents apply the LCF strategy. There is a 12.62% difference in the average normal welfare.

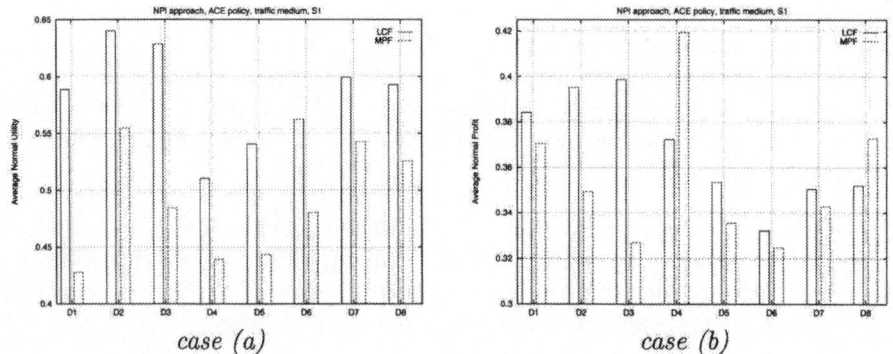

case (a)   case (b)

Figure 7.6: *Average normal utility (a) and average normal profit (b) obtained for the scenario S1 under medium traffic conditions for providers making use of different negotiation strategies.*

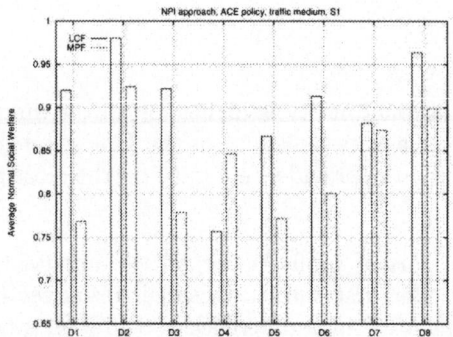

Figure 7.7: *Average normal social welfare observed for scenario S1 under medium traffic conditions for providers making use of different negotiation strategies.*

Interpretation   Gathering the experimental results for the three different scenarios reported here and considering the performance in terms of average acceptance rate and time discussed in Section 7.2.3, it is possible to draw some explanation:

- The average degradation in terms of utility, profit and social welfare, when passing from the LCF to the MPF strategy, is directly connected to the decrement of the average acceptance rate. With the LCF strategy, by reducing prices the probability that end users will accept the offers increases. In addition, because of the usage of less critical resources providers are able to allocate resources in a way that a good bandwidth connectivity level is maintained. This reduces the risk of future allocation failures. As a matter of fact, by examining the failure reasons in the MPF tests, a 50% of the allocation failures is due to the lack of intra-domain resources in some of the networks involved in the set up process. The remaining failures are mostly due to the fact that the prices proposed by providers are too high (30%) or the iterated negotiation process exceeds the time limits (20%). Notice that for the set of demands D4 and D8 the average normal profit is higher when providers adopt the MPF strategy (see Figure 7.6, case(b)) and this directly impact the relative performance of LCF and MPF in terms of social welfare. When examining more in details the simulation output files, this seems to be directly related to a higher number of risk averse end users on average, i.e., their average valuation of services is higher, therefore, they are willing to accept higher prices. For all the other set of demands there is a more balanced distribution (1/3 each) of risk neutral, averse and seeking end users. More investigation would be needed in order to estimate the influence of the end users valuation on average performance. This is discussed in more details in Section 8.3.

- The average degradation in terms of social welfare, when moving from the LCF to the MPF strategy, is increasing for larger scenarios. This is directly related to the observation made in Section 7.2.3. In larger networks, a potential higher number of interacting providers may have to cooperate for the allocation of a specific demand (i.e., longer end-to-end paths). This increases the potential conflicts that may arise and prevail upon the existence of a higher number of alternative paths that could be explored for the allocation of demands. This phenomenon has a stronger impact when providers offer higher prices on average (MPF approach) since the rejection rate increases. The average normal profit obtained in the different scenarios for the two negotiation strategies confirms that the performance become worse when provider agents make use of the MPF strategy in larger scenarios.

7.3.3 The Impact of Pricing Policies on Utility, Profit and Welfare

The way costs and prices are estimated in every network has a strong impact on network resources control and utilisation. Therefore, when providers make use of different pricing policies the multi-provider service set up performance can sensibly vary. In Section 7.2.4, the consequences of using either the ACE or the NOC pricing policies on average allocation time and on the average number of demands succesfully allocated have been examined. In the following, we consider more in details the effect on utility, profit and social welfare for the three network scenarios S1, S2 and S3.

All the results reported here have been obtained by averaging over 8 distinct test runs differing for the set of randomly generated demands D_i, with $i \in \{1, \ldots, 8\}$. Each set of demands has been generated under *medium* traffic conditions. For each set D_i the same network scenario has been populated once with agents applying the ACE policy and then with provider agents making use of the NOC policy. In both cases, NPAs make use of the LCF strategy. Figures 7.8 and 7.9 display the results obtained for the scenario S1.

Scenario S1

- U: The average normal utility is higher for all the different simulated sets of demands when the provider agents make use of the ACE policy. There is a 19.33% average decrease in U when the providers does not dynamically consider the criticalness factor for establishing prices (i.e., NOC policy).

- Π: The average normal profit is, for all the observed simulations, higher when providers dynamically price the criticalness. The ACE policy outperforms the NOC mechanism by a 10.04% difference on average.

- W: As a direct consequence of the two previous results, the average normal welfare is higher when the NPAs adopt the ACE policy. There is a 14.02% difference on average.

Scenario S2

- U: The average normal utility is higher when the provider agents make use of the ACE policy. There is a 20.5% average decrease in U when the providers use the NOC pricing policy.

- Π: The ACE policy outperforms the NOC policy. There is a 6.64% difference in π between the two.

- W: The average normal welfare is higher in all observed cases when the provider agents adopt the ACE policy. On the average, there is in this case a 12.64% difference.

Scenario S3

- U: The average normal utility is higher for all the different simulated sets of demands when the provider agents make use of the ACE policy. There is a 21.17% average decrease in U when the providers does not dynamically consider the criticalness factor for establishing prices (i.e., NOC policy).

- Π: The average normal profit is higher when providers dynamically price the criticalness. The ACE policy outperforms the NOC mechanism by a 6.33% difference on average.

- W: As a direct consequence of the two previous results, the average normal welfare is higher when the NPAs adopt the ACE policy. There is a 11.965% difference on average.

Interpretation Gathering the experimental results for the three different scenarios reported here and considering the performance in terms of average acceptance rate and time discussed in Section 7.2.3, it is possible to give the following explanations:

- The average normal utility, profit and social welfare are higher in all the simulated scenarios when the provider agents dynamically compute prices and costs (i.e., ACE policy). This is partly due to the higher number of demands succesfully allocated on average (as observed in Section 7.2.4), but also to the average higher benefit for both users and providers per allocated demand. There is an higher utility for end users since prices are on the average lower and a higher profit for providers since costs are on the average lower. With the ACE policy, this happens because costs are dynamically correlated to the current utilisation of resources and therefore prices can be reduced.

- The average utility of end users is higher with the ACE policy because less critical resources are cheaper than when pricing them with the NOC approach. Moreover, since when allocating non critical resources the opportunity costs of network resources drop down, also the net benefit of providers is higher on average with the ACE policy. On the other hand, since prices and opportunity costs of critical resources are higher when using the ACE policy, for heavier traffic conditions the degradation of the NOC policy with regard to the ACE approach is expected to decrease (see next item).

- Further tests of the same three scenarios for *strong* traffic conditions confirm that when increasing the traffic load, the average difference between the performance obtained for the two pricing policies decrease. Nevertheless, the ACE approach still outperforms the NOC policy.

  - The average normal utility is higher for the ACE policy, but the average differences for scenarios S1, S2 and S3 decrease to: 14.4%, 11.42% and 11.27%.
  - The average normal profit is higher for the ACE policy, but the average differences for scenarios S1, S2 and S3 decrease to: 9.03%, 3.99% and 1.06%.
  - The average normal social welfare is higher for the ACE policy, but the average differences for scenarios S1, S2 and S3 decrease to: 10.75%, 6.93% and 5.755%.

This indicates the necessity, especially for strong traffic conditions, of carefully adapting if necessary both the cost and the price functions in order, on one side, to optimise prices and, on the other side, to maintain a certain level of profit.

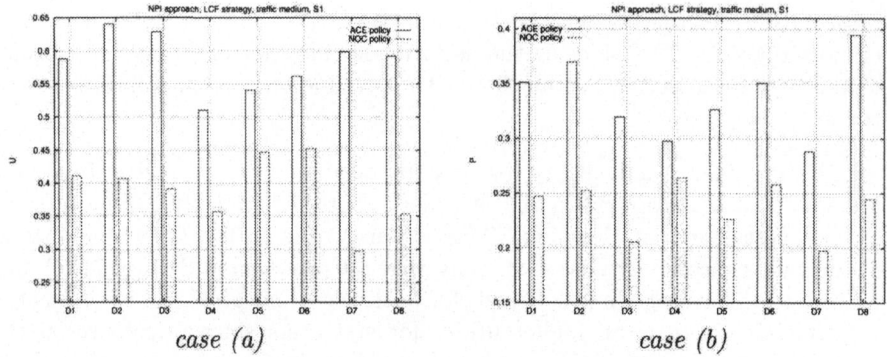

case (a)   *case (b)*

Figure 7.8: *Average normal utility (a) and average normal profit (b) obtained for the scenario S1 under medium traffic conditions for providers making use of different pricing policies.*

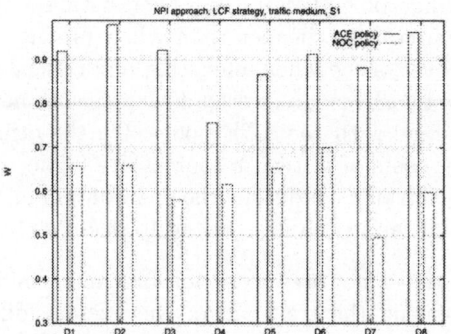

Figure 7.9: *Average normal welfare observed for the scenario S1 under medium traffic conditions for providers making use of different pricing policies.*

7.4   Comparing the NPI Approach with Alternative Coordination Mechanisms

The NPI approach has been compared with two main alternative automated paradigms for the coordination of peer providers: the FAS and the COB procedures.

- Within the FAS paradigm peer-to-peer providers coordination is regulated by static pre-existing agreements. Prices, available access nodes, guaranteed bandwidth and average end-to-end delay per service level are all fixed, therefore, there is no on-demand negotiation taking place between distinct operators.

	Approach	Contract	Rev. Distrib.	Confid.	Neg.
NPI	distr	dynamic	no	yes	yes
FAS	distr	static	no	yes	no
COB	centralized	dynamic	yes	no	no

Table 7.7: *This table summarises the main characteristics of the alternative coordination mechanisms discussed in this chapter.*

	S1		S2		S3	
Approach	Ar	Ta (sec)	Ar	Ta (sec)	Ar	Ta (sec)
FAS	92.01%	18.66*	90.31%	20.68*	84.42%	21.50*
NPI	94.20%	22.92	93.38%*	24.14	92.61%*	25.37
COB	94.42% *	24.49	92.08%	24.63	90.84%	24.72

Table 7.8: *Statistics on average acceptance rate and time for simulations of the scenarios S1, S2 and S3 under medium traffic conditions. The results are reported for three main distinct providers coordination methods: the NPI paradigm, the FAS solution and the COB approach. An asterisk indicates the best performance either in terms of Ar or in terms of Ta.*

- The COB solution relies upon the dynamic coordination of decentralised entities forming on-demand coalitions supervised, in the last phase of the service set up, by a central controller. The fundamental idea is that groups of providers join each other to coordinate their interactions by delegating the final part of the allocation process to a meta-NPA instead of running peer-to-peer negotiations. Every individual coalition is represented by a leader, which has the role of determining a global offer for the end user.

Table 7.7 shortly summarises the main characteristics of the three alternative coordination approaches presented in Chapter 6.

The experimental results reported in Tables 7.8 and 7.9 summarise the average performance obtained for the three distinct network scenarios S1, S2 and S3 for provider agents adopting alternative coordination mechanisms. The performance have been averaged over 20 distinct simulated test runs under *medium* traffic conditions.

7.4.1 The Networking Performance Comparison

The way providers coordinate their actions in order to set up services spanning multiple domains has a strong impact on both the average number of demands that can be succesfully satisfied and the average time that it takes to allocate these demands. In the following, the performance observed for the NPI, the FAS and the COB approaches are reported and commented.

- In scenarios S2 and S3, the NPI paradigm outperforms both the FAS and
  the COB approaches in terms of average acceptance rate. In scenario S1,
  the coalition-based approach shows slightly better performances. The FAS
  solution shows in all cases the worst performance.

  – *NPI versus FAS.* In comparison with the NPI solution, the most fun-
    damental limitation of the FAS approach is the poor flexibility of static
    inter-domain agreements to changing networking conditions. There is
    indeed no way to dynamically arrange peer-to-peer interactions unless
    the existing contracts are modified. In the NPI system, the organisa-
    tional structure of distinct provider agents can dynamically change to
    reflect the current state of the environment. The interactions between
    peer-providers are regulated by contracts and agreements, which are ne-
    gotiated on-demand. This increases the probability of allocating a larger
    volume of traffic with the NPI paradigm. This phenomenon is ampli-
    fied in scenarios with a higher number of interacting agents. Therefore,
    when passing from S1 to S2 and S3 the average observed differences in
    terms of Ar between the NPI and the FAS paradigm are respectively
    2.19%, 3.07% and 6.99%.

  – *NPI versus COB.* Dynamic peer provider negotiations are on aver-
    age more successful than coalitions supervised by a centralised meta-
    controller entity (at least in scenarios S2 and S3). This is mainly due to
    the additional costs that providers experience when joining a coalition.
    This includes the Clarke taxes and the extra communication costs in-
    curred for transferring information to the coalition leader. These extra
    expenses directly impact the average acceptance rate. On one hand,
    higher costs may not be covered by providers and, on the other hand,
    higher prices reduce the probability of finding end-to-end routes which
    satisfy end users requirements and reserve prices. While for a small
    number of interconnected domains, like in S1, it is still possible that
    the COB approach performs slightly better (notice that there is only a
    0.22% difference in Ar between the two paradigms), since the overhead
    and the cost of maintaining small coalitions is reduced, for larger multi-
    provider networks where the average number of providers per coalition
    increases, the NPI approach performs better. In scenario S2, NPI out-
    performs COB by a 1.3% margin and in scenario S3 there is a 1.77%
    difference in Ar. The main reasons explaining these kinds of results can
    be summarised as follows:

    * Increasing the number of networks does not necessarily increase the
      solution space.
    * There can be a potential increment of conflicts when the average
      number of interacting self-interested entities increases.
    * The adoption of a timeout influences the exploration of the solution
      space.

It is important to observe that, since network operators work at very high acceptance rate, i.e., the average acceptance rate Ar is close to 1, even small variations in terms of Ar can therefore be very significant in terms of performance.

- As expected, in all the observed tests the fastest approach is the static Fixed Agreements Solution.

  In comparison with the NPI paradigm, the average allocation time is lower with FAS since there are no dynamic negotiations taking place between distinct providers. The average differences in terms of Ta fall between 3.46 and 4.26 seconds.

  In comparison with the coalition based paradigm, Ta is lower since there is no need for exchanging information to delegate part of the set up process to a centralised meta-provider. The average differences in terms of Ta are included between 3.22 and 5.83 seconds.

- In scenarios S1 and S2, the average allocation time Ta is higher with the COB approach than with the NPI paradigm. This means that on average peer-to-peer negotiations are faster than coalition-based organisations. However, when the average number of providers involved in the allocation of a demand increases (scenario S3) coalitions converge to an agreement in a shorter time than negotiating agents. Notice that in this last case there is only 0.65 seconds difference.

- In terms of scalability, NPI outperforms both the FAS and the COB approaches. When passing from S1 to larger scenarios like S2 and S3 the average degradations in terms of acceptance rate are in fact lower in the NPI context.

  - In the NPI context, when passing from S1 to S2 there is a 0.98% decrease in Ar. This decrement goes up to 2.79% when passing from S1 to S3.

  - Within the FAS approach, when passing from S1 to S2 there is a 1.7% decrease in Ar. This decrease goes up to 7.59% when passing from S1 to S3.

  - For the COB solution, when passing from S1 to S2 there is a 3.34% decrease in Ar. This degradation goes up to 4.58% when passing from S1 to S3.

7.4.2 The Microeconomic Performance Comparison

The average results, which summarise the performance in terms of benefit of the three alternative approaches proposed for the coordination of distinct providers, are reported in Table 7.9.

Approach	S1			S2			S3		
	U	Π	W	U	Π	W	U	Π	W
FAS	0.3910	0.2327	0.6227	0.3820	0.2204	0.6126	0.2903	0.1832	0.4735
NPI*	0.5828	0.3373	0.9030	0.4994	0.3673	0.8520	0.4765	0.2680	0.7230
COB	0.5721	0.2847	0.8294	0.4873	0.3015	0.7806	0.4490	0.2391	0.6779

Table 7.9: *Statistics on average normal utility, profit and social welfare obtained when simulating three main distinct providers coordination methods: the NPI paradigm, the FAS solution and the COB approach. Experiments have been run for the scenarios S1, S2 and S3 under medium traffic conditions. A star indicates the best performance. In all the observed cases, the NPI paradigm outperforms both the FAS and the COB approaches.*

The NPI approach outperforms both the FAS and the COB solutions in all the different simulated scenarios (S1, S2 and S3) for all the observed test runs.

- *NPI versus FAS.* In comparison with the FAS paradigm, the capability of dynamically negotiating prices, resources and QoS characteristics determine higher benefits for both the end users and the providers. This is mainly due to a better utilisation and control of network resources which thereby determine average lower costs and prices, i.e., higher utility and higher profit. This phenomenon is also enforced by the capability, which providers relying upon the NPI coordination paradigm have, of allocating a higher number of service demands on average (see previous section for more details about *Ar*). In scenario S1, for instance, when passing from the FAS to the NPI solution, there is a 19.18% increase in the average normal utility, a 10% increment in the average normal profit and a 14.015% average gain in the normal social welfare.

- *NPI versus COB.* The experimental results confirm that the cost of organising providers coalitions has a strong impact on both the utility of the end users and the profit of the providers. In all the tests in fact the NPI paradigm outperforms the COB approach. Even when the average number of service demands succesfully allocated is higher with the COB paradigm (this happens in scenario S1, see Table 7.8), the average normal utility, the profit and the social welfare are higher for providers that dynamically negotiate prices, resources and QoS characteristics. The reason is that providers that individually negotiate with each other do not incur additional costs needed in the COB context for organising the coalitions, transmitting information to the centralised meta-controller and paying the taxes (see Section 6.3.3). In this way, costs are lower and prices can be optimised, i.e., higher utility and higher profit. In scenario S3, for instance, when passing from the COB to the NPI solution, there is a 2.75% increase in the average normal utility, a 2.89% increment in the average normal profit and a 2.255% average gain in the normal social welfare.

7.5   Results Summary

The most significant experimental results presented in this chapter can be summarised as follows:

- The NPI paradigm shows good scalability properties, at least up to a number of 12 distinct interconnected providers[5]. When passing from 4 to 8 distinct interconnected networks, for strong traffic conditions, there is a 1.26% maximum observed decrement in the average number of service demands Ar successfully allocated. In terms of average allocation time Ta there is a 1.29% maximum degradation (i.e., 0.33 seconds). When passing to 12 interconnected providers, for strong traffic conditions, there is a 2.18% maximum decrease in Ar and a 1.88% maximum increment in Ta (i.e., 0.48 seconds).

- The NPI approach shows promising performance when exploring the impact of different traffic patterns. Under strong traffic conditions it is still possible to succesfully allocate the 90.5% of received service demands within a 25.69 seconds on average.

- Within the NPI paradigm, the use of different negotiation strategies and pricing policies can significantly impact the performance.

    - The LCF strategy outperforms the MPF strategy. The key aspect is that when proposing the least critical resources first, the probability to succeed with the allocation of a specific demand increases because of a better utilisation of network resources and mean lower prices offered by providers to potential customers.

    - The ACE policy outperforms the NOC policy. This is directly related to the capability of the BI-based criticalness valuation of network resources, selected with the ACE policy, of dynamically estimating not only prices but also opportunity costs. This enables a better control of future demand allocations.

- When compared to the FAS and the COB approaches, the NPI paradigm has been observed to:

    - Scale up more effectively.

    - Allocate a higher number of service demands on average. This is even more evident for scenarios with a larger number of domains. Fixed agreements are unable to dynamically map the current network state and centrally supervised coalitions become too expensive and slow when the average number of providers per coalition increases.

---

[5]Notice that in realistic scenarios, the number of peer providers involved in the allocation of a specific end-to-end service demand is unlikely to involve more than 3 to 9 operators [77], [292], [72].

– Be faster than the COB approach and slower than the FAS solution. It is not surprising that NPI is slower than the static non-negotiation based FAS approach, but is interesting to observe that the service set up is faster on average when providers negotiate rather than when they form coalitions.

The absolute values obtained for the parameters observed should be considered as a tentative estimation of what could be effectively achieved in real networks. However, a more precise estimation would require (see also Section 9.3):

- More detailed data about real network configurations at both the inter- and the intra-domain levels. This also includes a more precise indication of the time required in every domain for retrieving the information needed from the management platform, or directly from individual network elements, in order to perform the various coordination steps at the agents level.

- More realistic traffic conditions. The traffic profiles considered in the experiments commented in this chapter have been generated by making use of the NPI Traffic Generator. The next step should be to run experiments with real traffic traces characterising the data exchange at the inter-domain level. Unfortunately, network and service providers are unlikely to reveal this kind of data mainly for strategic reasons.

- More realistic data about the utility, the profit, the pricing and the cost functions currently used by provider operators. Nowadays, however, given the highly dynamic and changing situation of the deregulated communication market, it is very difficult to characterise all the various components that influence the behaviour of providers and consumers.

- The relaxation of some simplifying assumptions considered in this framework to model the MuSS problem. This includes therefore the definition of a richer service model including, for instance, a more detailed description of service level agreements, the introduction of multi-cast and VBR demands.

Chapter 8

Discussion and Analysis

It does not matter what I believe,
it matters what I am able to prove.
– INSPECTOR DERRICK,

The initial part of this chapter introduces the main criteria that have been followed to analyse and assess the mechanisms developed for solving the MuSS problem. The first objective is to identify the key aspects to verify whether the NPI paradigm has the potential or not of properly coping with the fundamental challenges that providers face in the current deregulated communication market.

In the central part of the chapter, the analysis of the NPI paradigm and the discussion of the results obtained through experiments make it possible to underline its main strengths and weaknesses. In particular, various interesting properties emerge from the discussion of the relative performance of the alternative coordination approaches tested (namely the FAS and the COB paradigms, see Section 7.4). This critical and comparative analysis of the NPI paradigm also enables the extrapolation of the main open issues of this work. The focus here is on the aspects that, despite the relatively exhaustive results confirming the value of the NPI techniques still require more investigation.

The final part of the chapter discusses the choice of agents as the fundamental technology used for improving the way providers interact and coordinate their actions. This includes on one side an analysis of the major benefits that autonomous software entities endow and, on the other side, a discussion of the main risks that the choice of this technology brings in. This leads to the final part of the chapter in which the feasibility of the NPI solution is discussed with respect to several existing network technologies.

8.1   The Assessment Phase

In order to assess the paradigm proposed in this book for solving the MuSS problem, it is possible to consider several criteria. At the inter-domain level, for instance, negotiation mechanisms could be evaluated from a game theoretical perspective [266] by considering the following criteria[1]:

- *Efficiency*, which refers to the utility of having reached an agreement from a global point of view. 'Global' here refers to the totality of participants to the negotiation process. How this global utility is estimated can change. For instance, an agreement can be globally optimal when the social welfare, i.e., the sum of all agents payoff, is maximised. The social welfare maximisation principle is often used to estimate alternative mechanisms by comparing the solutions that these mechanisms lead to [282]. Another efficiency criterion is based upon the concept of Pareto efficiency. An agreement is Pareto-efficient when there is no other solution that makes at least one agent better off without making at least one other agent worse off [333]. Social welfare maximising solutions are a subset of the Pareto efficient ones. Once the sum of the payoffs is maximised, an individual agent payoff can increase only if another agent payoff decreases. In many real systems, however, individual self-motivated players may not be willing to nor capable of determining which choice is leading to the social welfare maximisation or to a globally Pareto efficient solution.

- *Stability*, which measures whether the designed mechanisms motivate each agent to behave in a particular desired manner or not. This also means that no agent should have incentive to deviate from a given strategy. In some frameworks, it is possible to design mechanisms with *dominant* strategies: an agent is best off by using a specific strategy no matter what strategies other agents use. However, often an agent best strategy depends on what strategies other agents choose, therefore, other stability criteria must be used. The most popular is the *Nash equilibrium* [209]. In Nash equilibrium, each agent chooses a strategy that is the best response to the other agents strategies. However, in some games no Nash equilibrium exists [166], and in some others there are multiple Nash equilibria, and it is not obvious which one an agent should play [198].

- *Simplicity / Complexity*, which evaluates how easy (tractable) / complex it is for participants to follow a given mechanism or protocol, in terms of both computational and communication overhead, and determine the optimal strategy.

- *Distribution*, which guarantees that the interaction mechanism or negotiation protocol is not centralised, e.g., that there is no single point of failure.

---

[1]More exhaustive details on assessment criteria from a game theoretical perspective can be found in [82], [281] and [150].

- *Symmetry* in the sense that the interaction protocol should not favour a priori one agent over another. This attribute is called *fairness* in the communication networks community [106].

Given the bounded rational nature of the agents developed in our framework, their self-interested character and the main requirements of multi-provider network, these classic criteria have been refined in the NPI context as presented in the following.

8.1.1   Valuation Criteria for the NPI Approach

The parameters used to assess the NPI procedures have been selected by considering the principal requirements in the current existing multi-provider scenarios[2] with the fundamental goal of identifying the main properties of the paradigm proposed here.

- **Efficiency** is estimated in the NPI context by making use of two types of indicators.

  - The *networking parameters* estimate the performance of the defined co-ordination techniques in terms of typical descriptors used in the communication network community, such as the average time required to allocate a service demand and the average number of demands successfully allocated [331]. The formal definition of these descriptors is given in Section 7.1.1.

  - The *microeconomic parameters* analyse the behaviour of end user and provider agents in terms of average utility, profit and social welfare. These descriptors have been formally introduced by the microeconomic theory [166] and refined in this volume as discussed in Section 7.3.

- **Scalability** can be considered as a property closely related to the efficiency of a system. More precisely, the idea is to estimate how efficiently the system performs when increasing its size. In the MuSS context, one of the most fundamental requirements is the definition of coordination mechanisms which perform well even when the number of different participants involved in the service allocation scales up (at least up to an upper bound which reflects real conditions of current environments[3]) and network topologies become more articulated.

- **Stability / Fairness**. The first characteristic establishes whether or not the rules for the providers coordination have been defined in a way that no agent

---

[2]See also Sections 1.1.4 and 2.3 for more details on current multi-provider networks requirements.

[3]Notice that in realistic scenarios, the number of peer providers involved in the allocation of a specific end-to-end service demand is unlikely to involve more than 3 to 9 operators [77], [292], [72].

has incentives to deviate from the decision making process specified by the NPI paradigm. When considering the second aspect, the objective is to make sure that the interaction protocol between peer providers and/or end users does not privilege a priori any of the participants.

- **Distribution**. In the MuSS context, the objective is the definition of a decentralised coordination mechanism reflecting the natural distribution of network resources owned and controlled by distinct providers.

- **Complexity / Completeness**. First of all, it is fundamental to define mechanisms that are feasible in terms of computational and communication overhead. Moreover, protocols, algorithms, strategies and all the techniques used should guarantee finding one solution eventually when solutions exist, and when there exists no solution, to find it out and terminate. These aspects strongly impact the efficiency of the system.

- **Flexibility** can be estimated in many different ways. According to the essential needs of multi-provider service environment [68], two fundamental elements have been identified:

  - The definition of a framework that is service independent. This means that even if the set of services considered in this book is inspired by a specific model (see Chapter 3), the NPI methodologies should be easily re-usable and/or refined when introducing different service features. This objective is fundamental in rapidly evolving communication networks, where new applications and services are appearing every day.

  - The definition of an architecture that should not limit a provider to a given business model. The goal here is to not prevent the possible introduction of additional economic mechanisms or business entities that might improve the performance of the NPI approach.

- **Openess / Interoperability**. Given the increasing number of actors and technologies in the current communication market, the ideal approach to multi-provider set up is an open and interoperable architecture. This implies the definition of a system which makes use of interfaces between distinct network domains that are standard and known.

- **Confidentiality**. The key aspect considered here is the definition of a co-ordination mechanism that allows internal and confidential data (such as network topologies, negotiation strategies, pricing mechanisms) not to be revealed to other providers.

- **Feasibility**. In this case, the idea is to estimate the possibility of adapting the NPI solution to existing and standardised network management and control frameworks (such as, for instance, the TMN architecture). This issues is discussed through a concrete example in Section 8.5.

Robustness and prevention of abuses are also fundamental properties of a good solution to the MuSS problem. Although security issues have not been addressed in this context, this aspect is discussed in Section 8.3.

8.2   Analysis and Discussion

Based on the properties identified in the previous section, we discuss the main strengths and weaknesses of the techniques proposed by this framework. A number of open issues directly arise from the relaxation of some the simplifying assumptions that have been made in the NPI context (see Chapter 3).

8.2.1   Efficiency

As anticipated in Section 7.1, the efficiency of the NPI paradigm has been evaluated by taking into account two main types of parameters. The *networking parameters* estimate the impact of automated multi-provider coordination mechanisms in terms of typical variables used to evaluate network performance [339]. This includes the average number of demands successfully allocated, i.e., acceptance rate, and the average time required for the allocation of a demand, i.e., allocation time. The *microeconomic parameters* describe the performance of the NPI approach in terms of end users utility, providers profit, and benefit to the society of participants to the service set up process measured by the social welfare. In this latter case, the basic idea is to evaluate the performance from an economic perspective.

¿From the conceptual description of the NPI paradigm, including the formal proofs characterising the various mechanisms discussed in Chapters 4 and 5, and the experimental results presented in Chapter 7 it can be seen that:

- *The NPI paradigm enables an efficient allocation of network capacity, i.e., bandwidth.*

  The reasons for that can be found in the combined application of powerful mechanisms at both the intra- and the inter-domain level:

  - At the intra-domain level the main novelty of the NPI approach is in combining *resource allocation* with *dynamic pricing* by making use of the Blocking Island (BI) formalism [93]. Within every single provider network, the possible routes are represented in a compact and flexible way in the Differentiated Blocking Island Hierarchy (Diff-BIH) representation of available resources (see Section 3.4.3). This structure is built and maintained by making use of the BI formalism, which enables the representation of bandwidth availability at different levels of abstraction in a very compact way. BIs highlight the *existence* and *location* of routes. Frei proves in [93] that there is at least one route satisfying the bandwidth requirement of an unallocated demand $d_u = (x, y, \beta_u)$ if and only if its endpoints x and y are in the same β_u-BI, where β_u is the required bandwidth. Moreover, all the links that could form part of such a route lie inside this β_u-BI. This principle can be used to quickly assess the existence of routes between end-points with a certain amount of available bandwidth, without having to explicitly search for such a route. Therefore, the Diff-BIH structure enable agents to speed up the decision making by accelerating the process of determining the intra-domain solutions space, i.e., the set of possible local routes that can support incoming service demands (see Section 4.3.2).

Furthermore, BIs are also used in order to estimate the costs and compute the prices that providers incur when allocating specific portions of bandwidth on network links (i.e., ACE policy). As shown in Section 7.2.4, this kind of approach outperforms traditional techniques which assume costs to be constant values that every provider internally establishes. This improvement is directly related to the capability of a BI-based policy of dynamically estimating not only prices, but also the opportunity costs that providers incur when reserving and allocating bandwidth on links. The ACE policy allows a better utilisation of network resources since the costs dynamically reflect the probability of blocking future incoming demands when making use of a certain amount of bandwidth on a link occupying a certain position in the Diff-BIH. The criticalness factor (see Definition 3.4.8) used for the cost computation estimates indeed the risk of isolating some network nodes when using bandwidth on a particular link[4].

– At the inter-domain level, a first strength of the NPI paradigm resides in the capability of creating organisational structures of distinct provider agents in a way that dynamically reflects the current state of the environment. The interactions between peer-providers are in fact regulated by contracts and agreements which are negotiated on-demand (see Chapter 5). This guarantees to control more effectively the way resources are allocated inside every provider domain, but also to make a more efficient use of the available bandwidth between distinct networks and satisfy changing interests of operators.

When compared to alternative coordination approaches, either based on the use of static service level agreements or upon the dynamic creation of providers coalitions supervised by a central meta-controller, the NPI paradigm shows a higher efficiency both in terms of networking and microeconomic performance (see Section 7.4). While fixed agreements are unable to dynamically map the current network state (FAS approach) and coalitions supervised by a meta-controller (COB approach) become too expensive and slow (especially when the average number of providers per coalition increases), automated negotiations combined with the use of powerful distributed constraint satisfaction techniques

---

[4]See Section 5.2.1 for more details on price and cost computation.

have the capability of better adapting the way resources are assigned to the incoming traffic demands.

- *The NPI mechanisms enable the participants to the service allocation process to achieve a level of profit satisfying their self-interests given the existing service and network constraints.*

The NPI approach outperforms both the FAS and the COB solutions in all the different simulated multi-provider scenarios for all the observed test runs. More precisely, given the existing constraints on resource consumption and conflicting providers interests, the social welfare is maximised as well as the average end users utility and providers profit (see Section 7.4.2).

In comparison with the FAS paradigm, the capability of dynamically negotiating prices, resources and QoS characteristics, produce higher benefits for both the end users and the providers. This is mainly due to a better utilisation and control of network resources which thereby determine average lower costs and prices, i.e., higher utility and higher profit. This phenomenon is also enforced by the capability, which providers relying upon the dynamic NPI coordination paradigm have, of allocating an average higher number of service demands than when static agreements limit peer-to-peer interactions.

The experimental results presented in Section 7.4.2 confirm that the cost of organising providers coalitions centrally supervised by a meta-NPA has a strong impact on both the utility of the end users and the profit of the providers. In all the observed tests, the NPI paradigm outperforms the COB approach. Even when the average number of service demands succesfully allocated is higher by adopting a centrally supervised organisation, the average normal utility, the profit and the social welfare are higher for providers dynamically negotiating prices, resources and QoS characteristics. The reason is that providers that individually negotiate with each other do not incur additional costs needed in the COB context for organising the coalitions, generating a meta-controller, centralising information and paying taxes (see Section 6.3.3).

A major weakness of the NPI paradigm that directly impacts both the networking and the microeconomic performance is in the *pre-reservation* mechanism of network resources. When a provider is negotiating with peer operators for the allocation of a given demand, in order to guarantee that an offer is still valid at the end of the negotiation it has to pre-reserve several resources in the network. This can strongly impact the allocation of other demands, which may have to be refused even if there exists in the network enough capacity to eventually accommodate new incoming demands. A key issue to reduce this negative effect is therefore the capability of agents to converge to agreements in the shortest possible time (i.e., short negotiations). This is why accelerating the decision making process of agents through the use of the Diff-BIH structure is a very important feature.

8.2.2   Scalability

The way providers coordinate their actions in order to set up services spanning multiple domains has a strong impact on both the average number of demands that can be succesfully satisfied and the average time that it takes to allocate these demands. The main idea here is to characterise how this impact varies when increasing the size of the multi-provider network. Larger scenarios imply that (1) the average number of providers involved in the allocation of a specific service demand increases and (2) the solution space that every provider has to handle becomes larger (as a consequence of a higher number of inter-domain links). Therefore, the following factors can deteriorate the performance of the interworking coordination:

- Along longer end-to-end paths (i.e., a higher number of domains) the probability that demands QoS requirements (such as bandwidth and end-to-end delay) cannot be satisfied increases.

- A higher number of potential conflicting self-interests increases the average duration of providers negotiations and the probability of not finding an agreement accommodating all negotiation participants.

- Demands allocated over paths involving a higher number of distinct networks have on the average higher prices. Therefore, end users may not afford to pay for the service to be allocated.

Despite these aspects, the experimental results demonstrate that the NPI paradigm scales well at least up to a number of interconnected network providers which reflects the possible working conditions in existing environments (see Section 7.2.1). The scalability of the NPI paradigm can be estimated more precisely by considering that:

- When passing from 4 to 8 interconnected domains, under strong traffic conditions, there is a 1.26% maximum observed decrement in the average number of service demands Ar successfully allocated. This means that it is still possible to allocate the 90.54% of received demands. In terms of average allocation time Ta there is a 1.29% maximum degradation (i.e., 0.33 seconds), which means a demand is allocated in 25.750 seconds on average.

- When passing to 12 interconnected providers, for strong traffic conditions, there is a 2.18% maximum decrease in Ar and a 1.88% maximum increment in Ta (i.e., 0.48 seconds). This means that is possible to allocate a 89.62% of received demands in 25.9 seconds on average.

Moreover, when compared to alternative interworking mechanisms such as the FAS and the COB approaches, the NPI paradigm shows much better scalability properties (see Section 7.4.1). This depends on the way the coordination of distinct providers is achieved in the distinct cases.

- The COB solution relies upon the dynamic coordination of decentralised entities forming on-demand coalitions supervised, in the last phase of the service set up, by a central controller. This implies additional costs (i.e., Clarke taxes) and communication overhead (transfers of information to the meta-controller). For this reason, in comparison with the NPI approach, centrally supervised coalitions become too expensive and slow, especially when the average number of providers per coalition increases.

- While the FAS procedure benefits from the main strengths of decentralised paradigms (see Section 6.5), in comparison with the NPI solution, the most fundamental limitation is the poor flexibility of static inter-domain agreements to changing network conditions. This aspect becomes even more critical in larger networks where the number of interacting providers increases.

In the NPI context, a provider is assumed to have a global and aggregated view of all the other operators in the scenario. This is necessary in order to perform the inter-domain source routing (see Section 4.3.2), i.e., compute global abstract paths along which the demand allocation process can be started. This presupposes that a periodic exchange of information takes place between the NPAs. This directly impacts the performance both in terms of scalability and efficiency. The intrinsic weakness of this kind of approach is in the need of maintaining an up to date and global multi-provider view, given the difficulty and the costs of exchanging information in a highly distributed environment. If providers are computing abstract path based on an obsolete view of the scenario the overall service set up process may fail. The main idea to improve the NPI paradigm is in this case suggested by the use of a BI-based hierarchical aggregation of information describing distinct networks in the multi-provider scenario (see Chapter 12 of [93]).

8.2.3 Stability

In order to evaluate how stable are the techniques proposed by this work, the fundamental objective here is to show that:

1. Given the interaction rules established by the NPI paradigm, agents cannot get higher profit than when they adopt a CSP-based decision making model.

2. Given the CSP-based decision making model agents cannot get higher profit than when they adopt the NPI coordination protocol.

1. Within the NPI paradigm, the rules that providers follow to interact are established by the Distributed Arc Consistency (DAC) algorithm. This relies upon the use of distributed constraint satisfaction techniques integrated within automated negotiation. From an individual provider perspective, therefore, it is possible to optimise the benefit only if (1) it can enumerate all internal alternative routes and compute the profit for each of them, (2) the mechanism for the selection of a specific intra-domain route does not prevent any of them from being

eventually chosen. While this latter aspect is guaranteed, in the NPI context, by the adoption of negotiation as the main route selection mechanism, the former issue is more critical. The crucial point here is that in real networks it is too expensive (and infeasible for bounded rational agents) to compute, represent, and store all the possible internal routes, including their price and cost. However, in the NPI paradigm, the combined use of the constraint satisfaction based decision making process with the BI formalism enable providers to handle an abstraction of all possible routes satisfying a demand given the current network state[5] (see Proposition 3.4.6). The blocking islands are used to verify if a given route exists before its actual computation and estimate its price and cost based on the current Diff-BIH.

2. From an individual and self-interested perspective, the profit can be optimised only if the protocol regulating providers interactions does not prevent any provider along the abstract end-to-end path from being able to select any of the feasible intra-domain routes. In particular, if the goal is to maximise the profit while preserving global bandwidth connectivity, a provider wants to be able to choose the least critical option inside its domain. The FAS and the COB approaches are typical examples of coordination protocols in which it may not be possible for an individual provider to arbitrarily select any of the feasible routes. In the first case, because of static agreements regulating prices and QoS characteristics that peer providers are expected to offer to each other, and, in the latter case, because individual self-interests are not necessarily satisfied when the coalition leader selects specific routes that maximise the coalition profit. On the other hand, when adopting the NPI coordination protocol it is possible that, once the infeasible solutions have been filtered out by means of the node and arc consistency steps (see Section 4.3.2), each provider arbitrarily negotiates its best choice.

Although there is no guarantee that for each provider involved in the service set up process the negotiation will terminate with the allocation of the most preferred resources, every operator is free to accept or not a certain deal that leads to an acceptable *good* rather than optimal profit. Notice that, if all peer providers negotiations terminate with an agreement about the most profitable solution for every participant, a Nash solution is found, [209]. This is indeed the only solution which satisfies the four Nash axioms:

- *Pareto efficiency.* If all peer providers negotiations terminate with an agreement about the most profitable solution s^*, there is obviously no other end-to-end solution that makes an agent better off.

- *Independence of irrelevant outcomes.* If all the irrelevant alternative end-to-end routes are excluded, but not the most profitable, then the most profitable

---

[5]Notice that interesting experimental results about the application of BIs for intra-domain demand allocation can be found in [93] and [352]. While [93] proposes an off-line centralised approach, the work discussed in [352] describes a decentralised and on-demand network control solution.

is still a solution. This is obvious as far as the needed network resources for **s*** are reserved in every network.

- *Symmetry.* The only options that are considered during the negotiation by the providers are the feasible outcomes, because node and arc consistency eliminate the infeasible ones.

- *Invariance under affine transformation.* The negotiation outcome does not depend on the scale of the profit function, i.e., the end-to-end route to which providers converge is the same, independently on the scale selected to associate numerical values to specific options. This is true in the NPI paradigm since there are non inter-domain comparisons in profit.

8.2.4 Fairness

- *Is the NPI paradigm privileging any of the specific participants to the multi-provider service set up process?*

By enabling peer-to-peer negotiations between the *initiator* NPA, i.e., the provider who first receives the service demand request, and the other providers involved along the abstract path selected for the service set up, the naturally privileged participant to the set up process is the *initiator* - as it happens in real settings. This is due to the fact that by being the final interlocutor of the end user this entity is responsible for both the abstract path selection, i.e., which peer providers will be involve, and the final global offer formulation including resources and fees for all the resources needed along the selected end-to-end global route. This process could be rendered more fair by enabling either a parallel search along the all possible abstract paths or by introducing a neutral centralised entity regulating all providers transactions. However, the former option is infeasible in real networks mainly because of its excessive operation overhead [45]. On the other hand ,the latter option may be in principle more fair, but concretely unable to satisfy individual self-interests. The NPI paradigm has been shown indeed to achieve better profits for individual negotiating agents than when adopting a neutral *supra-partes* entity like in the COB approach (see Section 7.4.2). Even though peer providers and end users can always decide whether to accept or not a deal, the dependency of the overall service setup process on the *initiator* provider can be considered as an intrinsic weakness of the NPI approach. It would be interesting to investigate the introduction of additional mechanisms that would allow a better control of the *initiator* from both the end users and the peer providers perspective.

8.2.5 Distribution

In the multi-provider set up environment, besides the natural distribution of network resources, information, control and interests, there are several fundamental reasons that promote a decentralised solving approach:

- Collecting information in a central point implies important communication costs including significant overheads and can become very inefficient or even infeasible for scalability reasons. This is experimentally confirmed by the relative performances of the NPI and the COB approaches (see Section 7.4).

- Centralised structures lack robustness, since any kind of failure damaging the centralised control unit compromises the whole system.

- In settings where different parts are self-motivated, collecting all the information into one agent is not necessarily feasible because of security and confidentiality reasons.

In the NPI system, provider agents solve the inter-domain allocation problem in a decentralised way. Every NPA assigns a part of the global route, namely the local part inside its network, and it negotiates with others in order to interconnect this route to form a global end-to-end connection.

- The distribution of control is achieved by delegating to distinct agents acting on behalf of distinct providers the supervision of their respective domains. This kind of organisation not only reflects the geographical distribution of resources, but also mirrors the self-interests of different operators.

- The natural distribution of information is guaranteed by the combined use of DCSP techniques and automated negotiations that allow the definition of consistent end-to-end routes without the need of centralising the data.

8.2.6 Completeness, Complexity and Soundness

When considering distributed constraint satisfaction techniques, resource allocation mechanisms and routing procedures another set of classic attributes for algorithm estimation such as *completeness, soundness, computational* complexity or number of *cycles, time* and *space* complexity [103], [7], [158], [370] are typically considered. In this context, the focus is on two main aspects:

- *Completeness* of the solving algorithms. An algorithm is said to be complete if it is guaranteed to find one solution eventually when solutions exist, and when there exists no solution, to find it out and terminate [370].

- *Computational complexity* indicates the number of steps or arithmetic operations or *cycles* required to solve the problem. A cycle corresponds to a series of agent actions, in which an agent recognises the state of the world, then decides its response to that state, and communicates its decisions [370].

    - *Time complexity* is the amount of time required for an algorithm (or agent) to find a solution (or to find that there are no solutions).

– *Space complexity* represents the amount of memory (or storage space) required by an algorithm and that generally varies with the size of the problem it is solving.

The complexity of the main different procedures involved in the solution of the MuSS problem has been already discussed in the previous chapters. More precisely:

- The complexity of maintaining the BIH decomposition of network resources has been considered in Section 3.4.2. Here, we just remind that the memory storage requirement of a BIH is bound by $O(rn^2)$, where r is the number of bandwidth requirement levels and n number of nodes in the provider network. This makes the use of these techniques computationally very cheap.

- The inter-domain routing complexity has been discussed in Section 4.3.2. In the NPI context, an *initiator* NPA computes on-demand, i.e., when the service demand arises, a list of abstract paths satisfying bandwidth requirements and minimising the delay. The complexity of the used in the NPI framework for the inter-domain route selection is bounded by $O(N^2)$, where N is the number of interconnected providers in the scenario). Notice that the number N of peer providers involved in the allocation of a specific end-to-end service demand is unlikely to involve more than 3 to 9 operators [72].

- The node and arc consistency procedures, performed for defining the set of feasible routes inside every domain, are discussed in Section 4.4. Their complexity is shown to be bound by $O(ab^6)$, where b is the maximal amount of boundary nodes (i.e., $b \ll n$) in the network of a provider along the abstract path $\mathcal{A}$ and a is the number of involved providers.

- The complexity incurred when following specific negotiation strategies is considered in Section 5.1.3. The time complexity of the LCF strategy, for instance, is limited by $O(n^2)$ where n is the number of nodes in the provider network.

An important aspect to underline when considering the complexity of the NPI paradigm is that both provider and end user agents are assumed to have (1) limited computational resources[6], i.e., a cost is associated to both computation and communication efforts, and (2) finite time to converge to an agreement, i.e., there are fixed negotiation timeouts. Therefore, the coordination mechanisms complexity (i.e., communication and time costs) is automatically taken into account in the experimental performance of the NPI techniques. Given the promising experimental results, it is possible thereby to say that the NPI paradigm is a computationally 'affordable' technique. In order to say more precisely *how* affordable this technique is in real networks, additional experiments need to be performed especially with more realistic data input characterising both the networks and the traffic.

---

[6]See Section 5.1 for more details.

To conclude this section, we remind that the completeness of the coordination mechanism adopted by provider agents in the NPI context has been formally proved in Chapter 5. The completeness is a direct consequence of how the overall DAC process has been structured in combination with (1) the use of routing mechanisms that are complete (see Section 4.3.2), and (2) the use of node and arc consistency that guarantee finding one solution eventually when solutions exist, and when there exists no solution, to find it out and terminate (see Section 4.4.2 for a formal prove).

8.2.7 Flexibility

The flexibility of the NPI approach can be assessed by considering two fundamental pressing needs of the current deregulated communication networks:

- The capability of flexibly dealing with an increasing number of heterogeneous services.

- The capability of flexibly interacting with an increasing number of new emerging business entities.

This leads to two main questions about the NPI paradigm.

- *To what extent the methods proposed by this volume are dependent on the beneath service model?*

The proposed coordination mechanisms rely upon the definition of a service model (SM) consisting of four main classes of service that differ for the QoS level they offer (see Section 3.2.4). This model has been assumed to be valid for all the interacting providers. According to that, a service demand is represented as a point-to-point connection requiring specific QoS guarantees (i.e., bandwidth and end-to-end delay) between two distinct network nodes. For this purpose, providers are required to be able of retrieving from the network (or, more precisely, from the management information base) the description of the current network state, in a way that enables to verify if these QoS requirements can be satisfied or not [36]. Moreover, a provider has to be able to eventually reserve resources in the network so that a given QoS level can be guaranteed. While this second aspect is directly related to the underlying network technology (see Section 8.5.1), the first task directly depends, in the NPI context, on the capability of building the Diff-BIH decomposition of available resources in the domain. Therefore, the only major dependency between the coordination methods proposed by this work and the SM underneath is the representation of service characteristics so that it is possible to map them into the Diff-BIH structure and vice-versa. More precisely, the NPI paradigm can be applied to any SM that makes explicit service requirements in terms of bandwidth.

- *Is the NPI approach constraining the adoption of a specific business model?*

The NPI framework can be considered as a virtual market place where different business entities can interact with each other. The main participants to this market that have been considered in this book are the final end users. i.e., *customers*, and the firms of communication services, i.e., *providers*. In this environment, the service demands spanning several domains origin, on one side, business-to-customers interactions between network providers and end users, and, on the other side, business-to-business interactions between distinct providers offering and buying services to each other.

Starting from this simple business model there are no specific restrictions on introducing other economic actors, while preserving the NPI paradigm as the main coordination mechanism for regulating peer providers interactions. For instance, it could be possible to add third trusted entities to which end users may delegate the negotiation process (or part of it) with network operators. Moreover, the vertical supply chain between end users and network operators may also involve service provider agents (SPAs), as suggested by the network management and provisioning scenario proposed by FIPA in [88] and considered in [83]. These SPAs would act as distinct business units able, for example, to act as mediators, brokers or matchmakers between end user and provider agents.

8.2.8 Openess, Interoperability and Confidentiality

In the current communication market, there is an increasing number of competing firms (service providers, networks operators, retailers, brokers, etc.), services and technologies that every provider has to be able to deal with. Despite an important effort of the networking community providers to converge to standard management and control solutions [325], [214], in many networks there are still several *legacy* interfaces and *ad hoc* components. In general, legacy systems are highly non-cooperative by nature and they typically make use of software that is proprietary to a single manufacturer. For this reason, in order to define an open and interoperable interworking framework two main requirements need to be satisfied:

- The use of standard information models and multi-provider management protocols.

- The wrapping up of ad hoc components and legacy interfaces so that standard mechanisms and protocols can work through the whole network.

In the NPI context, the first aspect has been addressed by adopting common agent-based interaction protocols, a standard agent communication language (ACL), a shared content language and a common ontology. The use of known interaction protocols establish the valid sequences of messages that agents can exchange in different phases of the service set up process. The use of a shared ACL enables the attitudes regarding the content of the exchange to be expressed by means of a common semantic. The ACL structure establishes, for instance, whether the content of the communication is, for instance, an assertion, a request or some form of query. Finally, a common content language and a shared ontology provide the syntactic and semantics support for the representation of the knowledge and information exchanged.

To address the complex task of wrapping up legacy components is out of the scope of this book. Furthermore, specific solutions would have to be defined based on the particular system to be wrapped. In the NPI context, the data describing the network state of the experimental scenarios is stored in files, which are dynamically updated during the simulation and which every NPA can directly access.

In conflict with the openness and interoperability need, there is an intrinsic exigency for self-interested and competitor entities to keep strategic information such as internal strategies, policies, utility functions or network topologies confidential.

- *What is the information that provider agents need to reveal to each other in order to achieve consistent end-to-end solutions?*

The multi-provider service set up represents a typical framework requiring the coordination of software entities that, despite self-interests, need to exhibit cooperative behaviour [179]. In the NPI context, the trade-off between individual objectives and coordination needs is reached by the combined use of distributed constraint satisfaction techniques and automated negotiation. On one side, NPAs can be considered *cooperative* since they exchange a minimal amount of information to find, within every separate network, the set of feasible local routes. This information corresponds to the boundary conditions reflecting QoS and connectivity constraints at the inter-domain level (see Section 3.3.1). Then, during subsequent negotiations to decide what specific route to select and at which price self interests prevail, i.e., every agent chooses the best local feasible route for itself.

8.3   Open Issues

The analysis of the NPI properties given in the previous sections identifies the main aspects which would require additional investigation.

Feasibility of Diff-BIH in Real Networks   One of the main strengths of the NPI approach comes from the application of BI-based techniques for both resource allocation and pricing inside distinct networks. Therefore, a critical aspect for the NPI paradigm to be effectively used is the feasibility of blocking islands based methods to real environments.

Very promising experimental results about the performance of BI based routing procedures have been presented by Frei in [93] and by Willmott in [353]. However, in Frei's framework BIs are used for off-line (i.e., not dynamic) demand allocation. A unique central entity is assumed to have a global view of all network resources and be responsible for allocating all incoming demands (that are assumed to be as well all known in advance). These assumptions clearly do not hold in the multi-provider service set up framework. Willmott proposes a distributed and dynamic version of Frei's approach in which the coordination of autonomous control systems linked together in an organisational framework has been used to efficiently solve on-line routing problems. All the network resources are in this case directly controlled by the same provider.

In the multi-provider context, despite the good empirical results reported in Chapter 7, an effective use of BIs methods would require considerable additional investigation with more realistic data characterising the network topologies, the data traffic, the services and the economic factors (such us utility, cost, pricing functions) existing in modern communication environments. This would also make it possible to verify the efficiency of the NPI paradigm for more complex and articulated intra-domain topologies.

Finer Control of QoS Characteristics Bandwidth is the primary QoS parameter for most (if not all) applications, but for many existing services (e.g., video-conference) this is not enough. Introducing more detailed service description (i.e., a richer service model), multi-cast demand and additional QoS parameters (i.e., relaxing Assumptions 3.2.4 and 3.2.5) would certainly have a significant impact not only on the NPI performance, but also on its feasibility. For instance, additive service characteristics, such as the end-to-end delay, need to be verified by summing all individual values describing route fragments forming the end-to-end path. In the NPI framework, this control is performed by the NPA initiator (see Section 4.3.2), assuming that up to date information about the average end-to-end delay values of global links and global nodes, which compose the path along which a demand is going to be allocated, is available. Thus, the feasibility of such a control is concretely conditioned by the availability of up to date information characterising domains where resources are distributed and dynamically change-able (i.e., their state). Even though some preliminary studies have focused on the integration of an NPI-like paradigm with a TMN network management system [36] more investigation and empirical tests would be needed.

More Realistic Cost and Price Computation In the NPI framework, cost and price determination has been enabled by the use of functions that take into ac-count the criticalness factor of network resources (see Chapter 5). Depending on the selected cost function, the same criticalness level can have a different cost and therefore determine different benefits for the providers. This not only influences the allocation of a single incoming demand, but also the global control of resources utilisation. In order to better estimate the efficiency of the pricing policy suggested by this work and the relationship between pricing and network control, it would be necessary to collect more exhaustive data about the various business entities populating the modern multi-provider market. This would include mode detailed information about the pricing policies and the profit models (when existing) cur-rently adopted by providers, retailers and buyers.

Multiple Network Provider Agents per Single Domain   A main assumption which the NPI paradigm relies upon presupposes the existence of at least one agent for every provider network (see Assumption 3.3.1). In all the experimental tests presented in Chapter 7, exactly one NPA was generated for each distinct network in the multi-provider scenario. However, for more complex inter- and intra-domain topologies, because of the number and the heterogeneous nature of the various tasks to accomplish for solving the MuSS problem, it could be necessary to introduce several NPAs. This may be beneficial even for simple topologies when the average number of service demands received by a provider increases.

Despite the potential improvement of performance, multiple NPAs per single domain would however have to coordinate themselves when reserving and allocating specific resources in the network. This could slow down the service set up process and therefore determine losses for providers either because the end users may go to faster competitors or because communications between the various NPA inside the provider domain would introduce too high costs. Therefore, additional investigation would be required for establishing the trade-off between the scalability and the efficiency of multiple NPAs per domain versus the complexity of coordinating them inside every network.

Different End Users Risk Attitudes   The risk attitude of an EUA is directly mapped into the way the service is estimated, which is summarised, in the NPI context, by the valuation function (see Section 5.1.2). In particular, given a generic valuation function, $\mathcal{V}(d,t) = p_r^u(d)\sigma(t)$, different users attitudes are specified by different shapes of the curve $\sigma(t)$. For a given demand d, the end user willingness to accept offers at a price closer to the reserve price p_r^u as the negotiation timeout approaches varies depending on the convexity degree of $\sigma(t)$. Therefore, for the allocation of the same service demand d, the same negotiation timeout θ_{neg} and the same reserve price p_r^u, at the same instant t the concession level of agents with different risk attitudes can be very different and determine quite dissimilar outcomes of the overall service set up process.

In all the experimental tests reported in Chapter 7, the risk attitude of the EUAs populating every scenario was randomly determined. The fundamental idea is that a random distribution of the risk attitude of potential customers more closely corresponds to real heterogeneous markets. Additional tests would be necessary to estimate more precisely the relationship between customers risk attitude and sellers pricing policies in different possible settings.

8.4   Choosing Agents

The idea of distributing communication network control and management tasks by making use of smart, cooperative and autonomous entities in network infrastructures has received considerable attention from both the DAI and the networking communities [343], [309], [194], [311], [351]. In particular, in [18], [235], [58], [33] and several contributions in [125], agents have been argued to have a strong potential for better solving various multi-provider tasks such as inter-domain connection configuration, routing, control, etc. In the following, the main strengths and risks of having chosen software agents for solving the MuSS problem are discussed.

8.4.1 Benefits of Autonomous Agents

The principal advantages of autonomous agents in multi-provider environments are closely related to the main properties these software entities are characterised by. As a matter of fact, these properties directly answer to the fundamental requirements envisaged for modern communication networks (see Sections 1.1.4 and 2.3).

- **Reactiveness**

  - *Agent Property.* Reactiveness refers to the capability of perceiving the environment and dynamically responding to changes that occur in it.

  - *Network Requirement.* From a network operator point of view, reactivity allows a quick response to many changing and dynamic situations in the network, including critical events such as alarms, fault and failures.

  - *Advantage*: Having distributed software agents running on (or close to) the elements to control, make it possible to quickly analyse and react to events, even before reporting them to human operators.

- **Goal Orientedness (pro-activeness)**

  - *Agent Property.* Pro-activeness is the ability to recognise opportunities and autonomously take the initiative. This is considered a direct consequence of internal goals rather than just a reaction to external stimuli.

  - *Network Requirement.* To administer large, complex and often heterogeneous environments require the delegation of tasks and objectives at different levels of the network management and control infrastructure.

  - *Advantage*: Pro-active software entities naturally represent an agency to which different policies and goals can be flexibly assigned throughout the whole network. Furthermore, software managers capable of taking the initiative can more easily prevent undesired situations to take place in the network.

- **Autonomy**

  - *Agent Property.* An agent is a system that can act without the direct intervention from others (humans or software processes) and has the control over its own internal state and actions.

- *Network Requirement.* For the large number of events that need to be processed at every instant at the network level, many network control and resource allocation tasks need to be automated, i.e., performed without the direct intervention of human operators. This is also true for a number of other actions such as resource pricing, service level agreements and contract negotiation.

- *Advantage*: Mobile or not, autonomous software agents can accelerate many networking and interworking operations without having to wait for external input or commands. This includes the coordination of peer providers acting either on behalf of human controllers or as an intelligent support for human-driven decisions.

These three properties together are also providing:

- *Strong adaptability* to network changes: agents have the capability to adapt themselves to unknown situations. If unexpected events occur in the networks agents would not stop working.

- *Robustness / Development and Reusability*: The distribution of control among specialised agents, which are able to handle specific tasks, enforce the flexibility and the robustness of the management approach. The tasks separation allows the development of distinct modules, that can be more rapidly created, modified or suppressed. A failure of a single distributed agent (or module) does not necessarily compromise the global architecture mechanism.

- *Flexibility / Easy Configuration* of policies: by defining agent goals and capabilities it is possible to define management and control policies at a higher level of abstractions with much richer semantics than having to manually configure network elements. Furthermore, in an agent-based framework, it is very easy to refer to different service models by simply sharing distinct and known ontologies.

- *Platform independence* of control strategies: agents can work on top of different underlying network elements and management platforms by abstracting from low level technical details to a more human-friendly and understandable form. A higher level of interaction facilitates the definition of common problem solving strategies which can be more easily applied given the current technological rate of change.

- *Scalability*: A distributed agent-based control system can be adopted to an increased network environment by adding new agents, and this does not necessarily affect the performance of the other pre-existing entities. Moreover, information describing different parts of the network is locally handled by distinct entities reducing the needs and costs of communication.

8.4.2 Impact and Risks of Software Agents

As observed by Wooldridge and Jennings in [357], in order to achieve the potential of agent technology in real systems, it is necessary to carefully assess the pragmatics of agent system development. In the following the aim is to identify the most relevant risks arising from the application of agents in the multi-provider context.

- *Trustability and security issues.* Because of the agents intrinsic autonomy and capability of acting without the need of external commands, their behaviour is not necessarily predictable. Therefore, trust becomes a fundamental issue. The true basis of trust, as Luhman discusses in [186], is uncertainty. In order to deal with an uncertain world and decide to act and pursue a specific goal without perfect knowledge and a stable environment, it is necessary to trust "enough" the available information, beliefs, actions, supports and other entities eventually involved in the environment. The fact that the agent-based computational paradigm creates an open and changeable world, which relies upon the interaction of autonomous entities, makes risks even more unpredictable and therefore trust even more crucial especially in multi-provider scenarios. In this context, distinct agents would act indeed on behalf of self-interested parties that are competitors in the current deregulated communication market scenario. Thus, in real environments agents holding strategic and private information would require a strong support in terms of security. Security can be considered as a set of mechanisms which reduce the risk of harm within the environment, i.e., security provides the fundamental building blocks for supporting the concept of trust. Despite the growing interest of the agent community on security issues [233], [234], this aspect still needs more work especially in multi-provider networks. Here, security mechanisms and policies are further constrained by the specific available management and control technologies.

- *Global Control.* Agents may have their own goals and objectives, which means that there is no direct control on what they are doing and how they are acting, but only on what they are supposed to achieve. When designing agents, it is indeed possible for developers to define agents beliefs, goals, plans and capabilities, but because of the agents ability to adapt themselves to changing environments it may become very difficult to control the impact of their actions in the environment they are embedded in. Given the layered and articulated network management and control infrastructure, the lack of direct control could affect the whole network performance. Moreover, an unpredictable behaviour of entities distributed in several places and affecting the network at different levels, could become very hard to detect, manage and correct.

- *Integration* with existing legacy system and existing network management and control platforms. In general, legacy systems are highly non-cooperative by nature and they typically make use of software that is proprietary to a single manufacturer. This leads to a large number of ad-hoc interfaces which need to be supported at different levels in current networks. When integrating an agent-based mechanism within these kinds of environments, it may therefore be necessary to define several specific wrappers so that data and operations at the network control level can be understood and executed by the agent world. This may become very expensive, especially in large and heterogeneous networks where several proprietary solutions usually co-exist.

On top of these main issues, the future application of agent technology in communication networks is also closely related to the possibility of overcoming the traditional separation between the DAI and the networking communities. Apart from the terminology problems already mentioned in Section 3.1, the division has created other obstacles:

- The continuing scepticism on the part of communications network engineers as to the utility and suitability (in terms of security, robustness, speed of operation, etc.) of agent technology. This has resulted in a lack of tested practical solutions and many approaches which have never made it beyond the test bed stage.

- The biggest obstacle for DAI researchers has perhaps been the technological complexity of the networks being studied. It would be fair to say that several of the promising methods developed by DAI researchers in the past have met with little success due to failings in the starting assumptions about the network domain.

- Agent solutions which have been proposed by the networking community have remained very simple and not leveraged some of the more powerful techniques developed by the DAI community.

8.5   Introducing NPI Agents in a Realistic Scenario

The application of the NPI approach to real network environments would require the integration of each agent in a real management infrastructure. In Figure 8.1, a possible concrete configuration is envisaged. In this example, the various NPI agents are integrated within a multi-provider architecture consisting of the two distinct networks A and B, which, following the TMN/TINA guidelines [140], [143], [323] interoperate at the management level through the use of a TMN-X compliant interface (see Section 2.2.1 for more details on TMN principles).

By interpreting users commands and input, each EUA is responsible for contacting providers so that services, eventually spanning different provider domains, can be automatically set up. In this example, let us assume that the EUA sitting in Berne, i.e., EUA 1, represents a global corporation willing to interconnect two

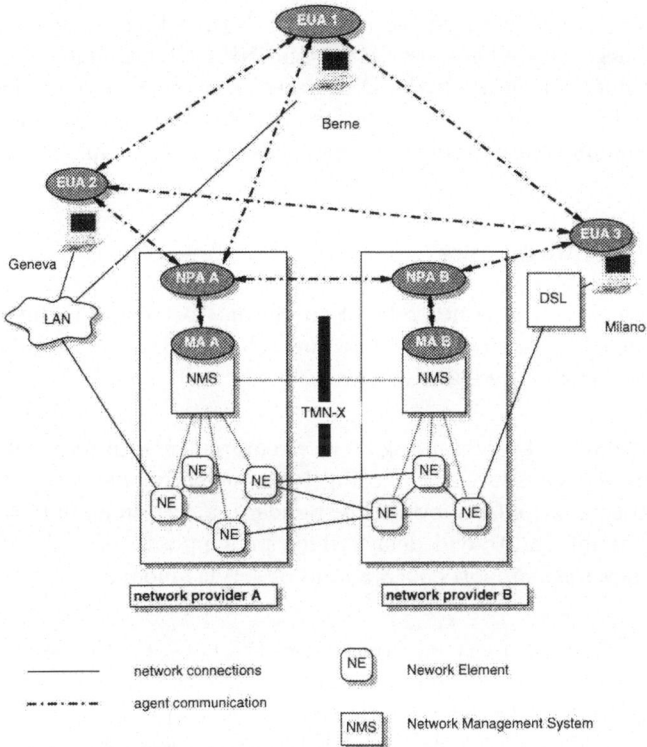

Figure 8.1: *This example shows the logical places where NPI agents could be integrated within a real TMN/TINA based multi-provider network scenario.*

geographically separated nodes belonging to its own virtual private network. The two remote nodes, Geneva and Milano, are directly connected respectively to network A and to network B[7]. The service set up process starts when the EUA 1 contacts, for instance, the NPA A asking for an end-to-end connection with the remote node in Milano. This entity firstly interacts with the Mask Agent (MA) A that act as a gateway between the network management system (NMS) (typically non-agent based) and the agent dimension. The main task of a MA is to convert and translate between low-level technical management primitives and information to a format that is understandable by agents, i.e., ACL based messages. Notice that for different kinds of underlying network technologies, ad hoc MAs (i.e., wrappers) would need to be implemented.

If inside the network A there are enough resources to support the request coming from EUA 1, the NPA A contacts the NPA B asking to collaborate and

---

[7]Remark that EUA 1 and EUA 2 are both directly interconnected by links forming a Local Area Network (LAN) that provider A supports.

initiate the DAC procedure (see Section 4.3.2). If the whole service set up process, including the negotiation between NPA A and NPA B, terminates successfully, an end-to-end connection between the DSL access controlled by provider B and the LAN controlled by provider A is established. At this point the various EUA agents can directly communicate exchanging messages between them.

8.5.1 The Requirements

Without considering the issues related to the use of agents in future networks (already discussed in Section 8.4.2), the main challenges for this vision to become true from a networking perspective are:

- The possibility of representing service requirements, in terms of media combination and QoS levels, according to the effective network availability. In the NPI context, a QoS level corresponds to a certain amount of bandwidth and a specific end-to-end delay, which presupposes an underlying network technology that supports such a kind of service model.

- The capability of reserving resources in the network: the network state can change quickly and a certain interval of time may elapse between the service demand call and its concrete allocation. Therefore, for effectively satisfying QoS requirements, providers need to reserve network resources.

With the traditional IP technology (IPv4) no QoS differentiation nor resource reservation can be supported. Therefore, the NPI approach is rather suitable for Internet environments integrating one of the following protocols: IPv6 [313], RSVP [245], MPLS [264], Tag Switching [258] and Internet Integrated Services [298]. Within these protocols, flow discrimination and priorities support the separate treatment of flows. In this way, demands for different QoS levels and explicit reservations on a per demand basis can be satisfied. Similarly, NPI could also apply to the Differentiated Services technology [182], which explicitly supports various categories of services that could be used to accommodate different QoS levels.

Finally, the NPI paradigm is directly applicable to ATM networks that support different classes of traffic with different QoS guarantees and embed mechanisms allowing the reservation of resources (see Section 2.1.1 for a more detailed overview of ATM networks). When an end node requests a connection with specific QoS requirements, PNNI [8] (the by default routing algorithm for ATM networks) is able to find a possible path (if any) satisfying the request and allocate the necessary resources in the network. Path selection and allocation is enabled by the Connection Admission Control (CAC) mechanism [273]. In that case, the NPAs would be responsible for directly controlling the CAC process.

8.5.2 The NPI Approach for Multi-Domain Configuration in Real Networks

Traditional multi-domain configuration management [140], [143], [323], [131] assumes that once a user has subscribed to a domain, he/she can request a connection that will be initiated by this domain, called the *originating domain*. The role of the originating domain may be different depending on the adopted multi-provider organisational model. The two most common traditional approaches are the *star* and the *cascade* models [66].

In the star organisational model, the originating domain or *initiator* provider directly cooperates with all involved peer networks in order to determine, maintain, and terminate the segments of end-to-end connections. On one hand, this model is quite rigid and only suitable for scenarios with a few networks. On the other hand, the advantage of this model is that it has fewer states and is therefore simpler to implement. In a pure cascade organisational context, each provider is responsible for the next segment of end-to-end connections. This model is more scalable than the star-based approach, but it can be quite complex to implement and control. Therefore, what seems more suitable for a scalable and flexible way of managing multi-provider configuration is a *hybrid* model [66], [155] combining strengths of both the star and the organisational approach.

The NPI paradigm can be considered as an hybrid approach since:

- *The originating domain or initiator provider represents the centre of a star-like organisation.* The overall multi-provider service setup is coordinated by the initiator provider that represents the final interlocutor of the end-user requiring the allocation of a specific service demand.

- *Peer providers interact with each other in a cascade fashion* for ensuring the consistency of the next segment of the end-to-end connection. During the distributed arc consistency phase, neighbour providers involved in the service demand allocation do exchange boundary constraints between each other without the need of centralising the process.

Chapter 9

Conclusions and Future Work

9.1 Scope of this Research

The recent liberalisation of the communication networks market is pushing network providers to evolve the way they interact with each other. Improving peer providers coordination is crucial in dealing with the increasing volume of traffic demands with stringent QoS requirements spanning domains under the control of different operators and current heterogeneity of network technologies. In parallel, the introduction of new services and the increasing number of diverse firms (service providers, networks operators, retailers, brokers, etc.) are intensifying the competition. Therefore, while improving the inter-domain coordination process, it is fundamental to ameliorate intra-domain resource allocation and control. In this book, this twofold issue is called the *Multi-provider Service Setup* (MuSS) problem.

At the inter-domain level, one of the most challenging aspects is the allocation of end-to-end routes guaranteeing specific QoS characteristics, given that the required resources are distributed and controlled by distinct self-interested entities that are unwilling to share strategic information. At the intra-domain level, the goal is not only to satisfy the QoS requirements of each incoming demand, but also to achieve global efficiency in the utilisation of resources so that providers profits can be maximised. Due to the complexity of these tasks at both the inter- and intra-domain levels, current solutions, which generally rely upon static agreements pre-negotiated by human network administrators communicating with each other by telephone or facsimile, are becoming more and more inadequate.

This approach proposes modelling a MuSS problem as a distributed constraint satisfaction problem for two main reasons. First of all, a distributed resource allocation problem is naturally represented as a DCSP. Second, this makes it possible to easily apply powerful DCSP techniques. These methods integrated with automated negotiation enable consistent end-to-end solutions to be found without having to reveal strategic information. In order to cope with the complexity of intra-domain tasks, such as resource control and pricing, distinct software agents acting on behalf of each provider make use of a dynamic abstraction technique called Blocking Island (BI), which clusters network resources according to their available bandwidth. This technique originally developed by Frei [93] has been refined and applied in the MuSS context enabling the definition of an efficient way of pricing network resources.

9.2   The Major Achievements

In general, the work presented in this book provides:

- A depth in in description of multi-provider interactions and service setup. This can be used to define a technology independent model of multi-provider interaction problems.

- An agent-based formalisation of such a framework and three different coordination models for self-interested entities: the Distributed Arc Consistency approach, the Fixed Agreements Solution and the Coalition Based Approach. The main benefit is in the definition of flexible and automated mechanisms that have the potential to support or eventually replace human operators for many low level aspects of network management and control.

- Negotiation instruments, such us protocols, strategies and reasoning models for bargaining agents acting on behalf of both providers and potential customers. The combined use of DCSP methods and microeconomic principles enable for an efficient and novel approach for the coordination of self-interested entities.

- A novel dynamic network pricing approach based on resource abstraction techniques that allow the evaluation of resource criticalness.

- An experimental analysis of the presented mechanisms for multi-provider network scenarios.

In the following, a detailed analysis of the most important contributions is divided in two parts. First, the major benefits of the NPI paradigm for multi-provider networks are given. Then, the significance of this work for agent-based coordination and negotiation frameworks is considered.

9.2.1 Multi-Provider Networks

A first group of contributions fall into the communication networks area: some aspects are directly relevant to inter-domain coordination and some other to intra-domain resource control.

An abstract (i.e., technology independent) model of the MuSS Problem The study of traditional approaches to multi-domain network configuration (such as the TMN and TINA paradigms) and the analysis of the most urgent and fundamental needs in modern communication networks enable a depth in description of multi-provider interactions and service setup (see Chapter 2). This allows the definition of a formal and technology independent model of the MuSS problem (see Chapter 3).

Based on this model, interactions between distinct operators are expressed as distributed constrained satisfaction problems. Software agents acting on behalf of different providers coordinate themselves by making use of powerful and extensively used DCSP techniques in combination with automated negotiation. Multi-provider resource allocation can be considered as a DCSP since local network resources (namely bandwidth) are distributed among agents and since constraints exist among them. More precisely, end-to-end routes are decomposed into fragments corresponding to independent decision makers (i.e., network provider agents). In DCSP terms, there is one variable (local path) per provider whose values are route fragments through that domain. Constraints between distinct variables ensure that route fragments connect and specific consistency methods are applied to rapidly prune out choices that cannot be part of an end-to-end route satisfying the demand QoS requirements. The space of feasible solutions, i.e., the set of possible intra-domain routes, provides the basis for subsequent negotiations. On top of this DCSP-based model, three main different approaches for providers coordination have been implemented and compared: the Distributed Arc Consistency (DAC) paradigm (Chapter 4), the Fixed Agreements Solution (FAS) and the COalition Based (COB) approach (see Chapter 6).

In the multi-domain configuration context, various works have previously addressed several issues proposing specific solutions (see Section 2.3.2). We strongly believe that the next step is the definition of a formal model of the MuSS problem that is independent on the underlying network technology (as far as this technology is consistent with the referred service model) has the potential to facilitate the development of standard techniques re-usable in a variety of concrete environments.

An Automated Framework for Provider-to-Provider Interactions Currently, many aspects of the interworking process are statically fixed by contracts (or service level agreements) and many steps of peer provider interactions highly rely upon human control. The NPI paradigm can be considered as a high level service which could be applied in communication networks in two different ways:

- In a short term period, as a smart support for human operators.

- In a long term perspective, as an autonomous system acting on behalf of humans and working at the connection management level.

The first version would help providers by computing, visualising and automatically verifying which local routes guarantee QoS requirements and satisfy inter-domain constraints, which directly derive from the rules, values and policies defined in the existing static contracts. In a future scenario, the NPI approach could supply an automated framework to route and negotiate the allocation of demands across distinct networks without the need for human intervention. Experimental results validate the defined agent-based paradigm and demonstrate the feasibility of the techniques developed in realistic network scenarios. Moreover, the comparison of the NPI-based coordination framework to non-automated solutions and/or automated approaches regulated by fixed contracts (i.e., where no dynamic negotiation takes place) confirms the benefits of automated and dynamic inter-domain coordination (see Chapter 7).

An intra-domain resource management approach automatically integrated with inter-domain coordination   The complexity of solving intra-domain resource allocation has been reduced by dynamically building abstractions that structure the problem space. In particular, the applied BI clustering scheme, developed by Frei and extensively discussed in [93], enables the representation of bandwidth availability inside every communication network at different levels of abstraction. This makes it possible to quickly assess the existence of routes between network nodes with a given amount of bandwidth, without having to explicitly search for such a route [94]. Therefore, in the NPI framework, the BI structure makes it possible to speed up negotiation decisions (see Section 4.3.2). The main novelty here is in the combined use of resource abstraction techniques for both resource control and pricing tasks. The use of a common data structure summarising bandwidth availability at different levels in the network in a compact form enable pricing to become a way of controlling and reducing resource congestion (see Chapter 5).

An Innovative Scheme for Pricing Network Resources   The main innovative aspects of the NPI pricing approach can be summarised as follows:

- At the *inter-domain charging structure* level, the fundamental idea is that end users are charged only by the network provider that they contact (i.e., initiator NPA). This unique global charge includes expenses for all different providers involved in the service allocation. This kind of approach is also called the *edge pricing* architecture [303]. In the NPI context, the main novelty, is that this kind of architecture is implemented without binding rules on how prices are established in different networks. The main advantage, from the provider point of view, is therefore that the local pricing process can be performed by every operator indipendentely of how others decide to estimate their own resources. Moreover, on-demand peer agreements allow the computation of prices that dynamically reflect the current criticalness of the resources selected along the selected abstract path. From the users perspective, this approach simplifies the billing and accounting procedure. There is indeed only a bilateral negotiation process and a unique bilateral contract between the end user and the initiator NPA. Although, this inter-domain charging structure requires extra communication between distinct agents for the dynamic peer-to-peer coordination, experimental results validate the potential of this approach (see Chapter 7).

- At the *intra-domain level*, the fundamental idea beyond the pricing model developed in the NPI context is the dynamic computation of prices and costs by considering the current network resource criticalness. The first novelty of this approach is in the way the criticalness is estimated. Given the Diff-BI decomposition of the network (see Section 3.4.3), if a link is critical it has few available resources (i.e., the price/cost increases). However, if a link has few available resources it is not necessarily critical (i.e., not necessarily expensive). This happens, for instance, when a link l with low available bandwidth β_x is clustered in a β-BI in which there exist alternative routes guaranteeing the interconnection of the end-point of l with a bandwidth of $\beta > \beta_x$ (see Chapter 3).

  The computed service prices consist of two main fees. A *usage based* component ensures that end users have incentive in declaring their true QoS requirements. A *cost-based* component allows the increment of prices/costs to control the network congestion. The second main novelty is directly related to the dynamic cost computation. In many current communication networks, costs are assumed to be constant values that cannot be directly related to the current resource consumption (i.e., NOC policy, see Section 5.2.1). Therefore, in order to achieve a certain level of profit providers may not be able to optimise the prices. This increases the risk of loosing potential customers. A more effective way of evaluating costs would be more beneficial for both providers and users. Experimental results demonstrate that, in comparison with static costs, both the end users utility and the providers profit as well as the number of service demands successfully allocated increase when the costs are dynamically evaluated with regard to the current network state (see Sections 7.2.4 and 7.3.3).

While the inter-domain charging structure establish the way providers interact to finally make proposals to potential customers, the intra-domain pricing paradigm represents a possible *optional* approach. This means that, in the NPI system, distinct NPAs making use of different pricing policies inside their domains can coexist and properly interoperate. The ACE paradigm and the NOC approach are the two main different intra-domain pricing policies implemented in the NPI context (see Chapter 5).

A Prototype of an Open Electronic Market for IP and Telecom Services Negotiation and market-based techniques for multi-provider service provisioning can facilitate and promote the integration of activities for the on-line trading of IP and Telecom goods with traditional network management functionalities such as planning, configuration, resource allocation, monitoring, etc. Virtual market-places facilitate commerce transactions by bringing together a high number of potential buyers and sellers [168]. Many service and network providers are therefore considering, and in some cases already using[1], electronic market places to offer services. However, even though this kind of business is becoming popular, the risk of failing or not being effective is not negligible. For this reason, the development of prototypes is essential for better evaluating the potential impact, risks, limitations and advantages of using computational markets. The multi-provider networks simulator, which has been developed to validate the NPI approach, by means of common and standardised agent communication facilities, supplies an open and interoperable testbed. Standard mechanisms facilitate access to electronic service negotiations to a larger number of potential customers and/or sellers of Telecom and Internet services. Moreover, the usage of high level agent communication languages (based on speech act theory [289]) instead of low levels primitives such as SNMP or CMIP routines makes it possible to abstract from technical networking details and to enhance agents conversations with a stronger semantic foundation behind the messages sequence.

Furthermore, the multi-provider networks simulator has the potential to be modified and re-used for studying other aspects of networking and interworking processes.

Exploration of computational issues in multi-provider resource allocation The experimental analysis and the discussion of the potential benefits and the main limitations of automated solutions in realistic network scenarios represent an important first step toward the effective use of distributed software control systems acting without the direct intervention of human operators.

The comparative advantages of automated negotiations and intra-domain dynamic management of resources, including pricing, are examined (see Chapters 7 and 8). One of the main strengths of the NPI paradigm is the capability of supplying a rapid and efficient answer to the MuSS problem, without the need for different operators to reveal a complete description of the internal topology and confidential information. The data to be exchanged concerns the set of possible access points

---

[1] As reported by Makris, at least seven bandwidth brokers are already active on the Web [195].

that a network provider can use for a specific demand. It is reasonable to assume that providers which need to interoperate must exchange a minimal amount of information, such as a topology aggregated view [212], [102], [145]. Experimental results show that the performance of automated solutions depend heavily not only on network characteristics (i.e., topologies, network resources, such as bandwidth and nodes, etc.), but also on negotiation strategies and pricing policies (see Sections 7.2.3 and 7.2.4). This confirms that traditional myopic approaches to the interworking problem focusing only on resource allocation and network management issues need to be improved. Automated and standardised information exchange between software agents compared to the exchange of possibly ambiguous facsimiles, telephone calls and sometimes unreliable e-mails between humans, seems a valid argument for relying on software mediators.

9.2.2 Agent-Based Coordination and Negotiations

Negotiation has been proved to be a powerful instrument for coordinating both self-motivated and/or cooperative agents. In the NPI framework, the focus is on the interactions of *bounded rational* entities that despite self-interests need to cooperate. The negotiation approach proposed in this volume is a heuristic model that falls into a particular category of bounded rationality where boundaries, limitations and constraints are intrinsic in the CSP-based decision making model that the developed agents rely upon. Within this context, the main contributions of this work can be summarised as follows.

An Approach for the Dynamic Creation of Self-Interested Agent Coalitions In general, self-motivated agents choose the best strategy for themselves, and this cannot be explicitly imposed from outside. However, in several contexts it may be necessary or just more profitable for groups of self-interested entities to cooperate and form coalitions. In this context, despite competitive interests, providers need to act in a cooperative way, since they need to exchange a minimal amount of information to finally converge to consistent end-to-end routes for the allocation of multi-provider service demands. The DCSP based interaction techniques proposed in this can be considered as an innovative approach for the dynamic formation of agents coalitions in relation to the current state of the environment. Providers interests and policies in addition to the network state represent the existing constraints on the way peer-providers interactions can take place. While in the NPI context there is an implicit creation of dynamic and decentralised coalitions, in Chapter 6 an alternative mechanism for centrally supervised providers coalition is also discussed. From the experimental results reported in Chapter 7 the latter approach enables for better performance.

Although in this book the main focus is on the automated coordination of software entities for solving the MuSS problem, the dynamic formation of coalitions is becoming a key aspect in several fields. This is particularly evident in modern on-line business environments, where there is an increasing number of virtual enterprises. The techniques proposed in this volume are expected to be possibly adapted and re-used for addressing the coordination needs for self-interested and constrained entities forming these virtual organisations.

A Heuristic Model of Bounded Rationality based on a CSP Decision Making Approach   The negotiation framework developed within this context is a heuristic model that falls into a particular category of bounded rationality [269] where boundaries, limitations and constraints are intrinsic in the CSP-based decision making model that the agents rely upon. First of all, NPI agents have limited information about their environment, including other entities populating the multi-provider scenario. In addition, both end user and provider agents have limited computational resources (i.e., a cost is associated to both computation and communication efforts) and a finite interval of time to converge to an agreement (i.e., fixed negotiation timeouts). Two main alternative strategies relying upon an innovative way of estimating resource criticalness in the network have been proposed (as discussed in Chapter 5) and experimentally validated (see Chapter 7). Furthermore, the combination of the NPI coordination paradigm with the CSP-based decision making model has been shown (in Chapter 8) to have the potential of determining a Nash solution [209].

An important aspect is that, since constraint satisfaction methods are domain independent, they have the potential to be re-used in other application fields. Therefore, the CSP-based decision making model can be considered as a generic approach that can be adapted to other problems requiring the coordination of self-interested and bounded rational entities.

Reusable Service Models and Ontologies for Agents   The defined service model, content language and ontology for IP-bandwidth and Telecom services, including VPN provisioning, have the potential to be reused in other agent-based frameworks. In developing service abstractions and the related ontology, a particular effort has been devoted to defining models that are both *realistic*, i.e., based on existing networking frameworks, and *modular*, i.e., for easy assembly, composition and integration with other services and/or ontologies. In particular, the agent communication facilities proposed by this work have been re-used in the context of the MACH Project[2] for the implementation of a multi-agent auction house selling IP-based services [34].

A Model of Bilateral Negotiations in a Standard Framework   The majority of more traditional multi-agent systems have been focusing on interactions between one-to-many (e.g., auctions) or many-to-many (e.g., markets) participants. Few recent frameworks (as discussed in Section 2.3.1) have been focusing on bilateral (i.e., one-to-one) negotiation between autonomous agents. The approach developed

---

[2]More details about the MACH Project can be found at:
http://liawww.epfl.ch/~calisti/MACH

within this book models bilateral negotiations and provides a framework where standard communication facilities (including standard negotiation protocols) define an open market accessible to agents developed in different frameworks.

9.3   Directions for Further Work

9.3.1   Summary of the Open Issues

As discussed in Section 8.3, despite the relatively exhaustive results confirming the value of the NPI techniques for solving the MuSS problem, there are a number of issues that still require more investigation. More precisely, further work would be needed in order to better address the following aspects:

- *The feasibility of the Diff-BIH in real networks.*

- *A finer control of QoS characteristics.*

- *More realistic cost and price computation.*

- *Multiple network provider agents per single domain.*

- *A more depth in analysis of different end users risk attitudes.*

- *Integration of the NPI paradigm with legacy systems and existing non-agent based network management and control platforms.*

- *Introduction of more sophisticated security mechanisms.*

In addition to a more depth of investigation of these open issues, several directions for future work are suggested by the potential application and adaptation of the NPI techniques to other problems in various fields.

9.3.2   Other Applications

Besides the specific problems related to the multi-provider service set up process, the work described in this volume is relevant to other frameworks requiring coordination of software entities that, despite local constraints, need to exhibit cooperative behaviour to *"improve their local performance"* [179].

Service Management   In this volume, the focus is on the service setup process that represents the initial phase of the service life-cycle during which the necessary resources to satisfy the end users QoS requirements are negotiated, selected and configured. The NPI paradigm could be extended in order to automatically support also the service maintenance (or management) during its whole life-cycle.

A major issue in current service management is the capability for providers to offer *reliable* services. The service should be reliable in the sense that the agreed quality of service is met during the whole (or during a certain pre-fixed percentage of its) duration, and the risk of unexpected termination of the service is minimised. Furthermore, the service should be *robust* in the sense that it can recover from most exceptions that can happen in the networks. For instance, when a link that is part of a certain connection cannot be longer provided because of a hardware fault in a switch node, the provider responsible for that link should be able to automatically set up an alternative connection to keep the service alive (i.e., fault management).

In order to satisfy reliability and robustness needs, providers have to make use of flexible, dynamic and efficient mechanisms for managing and control the resources in each network involved in the service provisioning. Network monitoring and fault management imply additional costs that can determine variations in the offered QoS level. This may require the re-negotiation of prices and service characteristics with both the final end users and the other peer providers involved in the demand allocation.

The application of the Diff-BIH network decomposition in combination with the ACE policy seems particularly suitable for this task, since it would enable NPAs to dynamically verify the current network state and eventually recompute costs and re-negotiate prices. An additional feature that an automated NPI-like multi-provider framework would offer is the capability of supplying mechanisms for a efficient fault management approach through the use of DCSP based methods. The need to exchange fault management information across domains can be a problem because this type of information may be seen by a provider as highly sensitive. If this information can be expressed as a set of constraints on boundary network resources, arc consistency techniques enable distinct providers to dynamically react to 'faults' without the need to reveal the causes of these faults to each other.

Dynamic Service Level Agreements   Currently, many aspects of the interactions between network providers are regulated by static long-term Service Level Agreements [331] (SLAs) and/or contracts that define the number and available capacity of links and network nodes (or access points) connecting one network domain to another, the access nodes, their prices, etc., without taking into account the current network state. As a main consequence, intra-domain tasks and performance are strongly constrained, since the possibility to dynamically accommodate prices and balance the load on network resources is very limited or even not possible. This generates inefficient network resources utilisation with consequent profit losses [111].

Within the Differentiated Service (DS) framework [297], *bandwidth brokers* have been proposed to implement an automated SLAs exchange according to the current network state [212]. Although these entities provide a certain degree of flexibility one of the most debated issues is the capability of effectively combining network control and resource pricing (see Section 2.2.2).

The network provider agents developed in the NPI framework can be considered as bandwidth brokers that effectively combine resource control and pricing by making use of BI-based techniques. Additionally, the distributed constraint satisfaction based formalism could be directly applied for verifying which consistent SLAs can be established between distinct domains by automatically taking into account both service and network constraints. *Service constraints* correspond to QoS and connectivity requirements as expressed by a specific service demand. *Network constraints* are imposed by network resource availability, providers control and/or management policies [111], [240].

Active Networks Active networking is an emergent paradigm for building current network architectures based on the new capabilities of IPv6 [313]. Active networks provide the means to insert control processes and decision logic into individual nodes such as routers or switches. This is enabling the use of more intelligent and flexible network management schemes [322]. However, for these active approaches to be effectively used in real networks a crucial aspect is the coordination of the different programs running on distributed active nodes.

The potential of integrating autonomous entities within active nodes inside a specific domain has been discussed, for instance, in [354] and [124]. Having distributed software agents running on (or close to) the elements to control, make it possible to quickly analyse and react to events, even before reporting them to human operators. In this context, the NPI paradigm has the potential to be applied as a paradigm for achieving an automated coordination of active nodes clustered in networks controlled by different authorities. This would require the integration of NPI agents within two main specific types of active nodes. In the case of an ATM network, for instance, the network provider agents would be directly responsible for the on-demand *Call Admission Control* (CAC) process. In packet networks like the Internet, the NPAs would be responsible, for example, for the *per-flow* traffic control and resource reservation.

Federation of Traders in Electronic Markets The multi-agent technology is expected to play a fundamental role in changing the relationship between consumers and service providers as well as the relationship between distinct business entities offering services to each other. Through the use of autonomous software assistants, customers would have a more flexible control and a better access to a wider range of services. On the other hand, providers would be able to dynamically interact between each other for aggregating existing services, adding new facilities, and generating completely new service packages. Currently, many enterprises rely on a 'self-service' customer model. For instance, today, a specific end user can buy a digital camera accessing a Web interface where he/she is expected to manually enter personal data and press the right button. Tomorrow, electronic personal assistants may interact with providers to create a tailored solution based on specific consumers requirements and preferences. In parallel, provider agents could offer a broader range of services due to greater dynamicity in combining services from different network operators. However, two main key issues for this vision to become true rely upon the capability of agents to:

- Dynamically coordinate each others activities for aggregating, buying and offering services.

- Effectively manage the information required for agent decisions to be effective in this complex and heterogeneous environment.

The combination of automated negotiation and constrained satisfaction techniques has the potential of efficiently addressing these aspects by marrying two important characteristics:

- Automated negotiations provide methods for the coordination of both self-interested and/or cooperative entities [150].

- Constraint satisfaction

  - Provides search algorithms and consistency techniques which are both very simple and compact to implement, and at the same time enable complex behaviours with reasonable efficiency [191].

  - Can represent complex information in a compact form [338].

The techniques proposed in this context could hence be adapted and re-used to address both the coordination needs for self-interested and constrained entities forming virtual multi-provider enterprises (i.e., B2B relationships) and the interactions between customers and service providers (i.e., B2C relationships).

9.4   Final Conclusions

The dynamic coordination of self-interested entities in complex and changeable environments such as today's multi-provider communications networks is a very difficult problem that is receiving increasing attention. This work argues that:

- The combined use of DCSP techniques and automated negotiation can be the basis for efficient peer provider coordination and flexible interactions between end users and providers.

- The use of dynamic abstraction techniques can simplify and improve network control, including resources allocation and pricing.

The mechanisms developed to model and solve the MuSS problem provide an automated, flexible and natural way of addressing many urgent, fundamental needs in modern networks that human operators are more and more incapable of dealing with. Experimental results demonstrate the potential performance of these mechanisms for a number of simulated scenarios. Even though experiments with more realistic traffic and networks data would be needed in order to better estimate the impact of the NPI paradigm in real environments, we strongly believe that this kind of automated and intelligent approach will play a key role in the future of communication networks and applications that directly build on top of them. This may be the case of systems providing services such as e-banking, e-financing, air traffic control, production planning and control in the process industry.

We therefore argue that several complex problems requiring coordinated interactions of distinct self-interested entities in highly dynamic environments could benefit from the use of distributed constraint satisfaction techniques combined with automated negotiation. The paradigm proposed in this volume can hopefully be considered as an initial step into that direction.

Appendix A

Pricing in Communication Networks

A.1 A Short Background on Pricing Networks

Pricing network resources for services that span domains owned by different authorities, either Telecom or Internet domains, is a very complex task that received an increasing attention over the last ten years. In this section, a short background on the main existing pricing approaches is given [1]. The main purpose is to underline the fundamental challenges that have to be faced when designing an architecture for pricing network resources. This mainly consists of:

- Defining what are the price components, i.e., access, setup, usage fees, identifying therefore the information needed to compute the prices.

- Developing efficient mechanisms to collect this information, keep it up to date and finally compute the prices.

- Defining what are the main objectives (e.g., congestion control, profit maximisation) when pricing resources.

These crucial issues all derive by the need of finding an appropriate trade-off between networking efficiency (including technological feasibility) and economic objectives. This objective is even more complex to achieve in multi-provider networks, since several self-interested entities can apply different pricing policies.

A.1.1 Price Components

Prices of communication services can include three main basic components:

---

[1]For a more complete overview we refer to good surveys given in [257], [24], [306], [237]

- A *flat-rate* is usually collected as a monthly (or annually) charge for using a specific access connection to the network. The price depends on the capacity (e.g., average throughput) of the connection.

- A *usage-based* fee is a component that quantifies the current resource usage in terms of time, volume or QoS. This can be computed by deploying a specific *price function*.

- A *content* dependent fee can be added depending on the specific content supplied with the service (e.g., video-on-demand [314]).

Traditional Internet services for final end users are usually priced on a flat-rate basis: the only considered price components is a fixed access charge. This is also indicated as an *access-based* pricing scheme. Since no accounting is required and the effort for billing can be limited to a minimum, this approach is very easy to implement. The main limitation of flat-rate pricing emerges during congestion. In this case, the marginal cost of forwarding a packet is not zero, but flat pricing does not offer any (dis)-incentive for users to adjust their demand. Although it has been argued that flat pricing and over-provisioning is still feasible given the decreasing cost of bandwidth and available time for network provisioning [218], there exist several cases in which adding bandwidth is not technically available or fast enough.

As a more responsive alternative, dynamic pricing schemes taking into account the current state of the network have been increasingly proposed. Usage-based pricing helps in controlling and reducing resource congestion if higher per volume prices are introduced when the network is more loaded. In addition, usage-sensitive pricing can be used as a mechanism to reinforce the fulfilment of different QoS requirements imposed by different types of services, applications and users.

In [49], Clark proposes pricing as a way to offer different options in network usage to the users in a controlled manner. The idea is to implement a charging scheme based on the *expected capacity profile* that is a sort of traffic contract negotiated between users and providers. The preliminary observation is that flat-rate pricing, although it is very simple to implement and encourages usage, does not reflect congestion cost. On the other hand, usage-based pricing schemes, can discourage 'heavy' users and lead the providers to increase prices to recuperate costs, but have the advantage of controlling network congestion. Thus, the expected capacity approach has been designed with the aim of capturing the advantages of both schemes. Users are charged only on the expected capacity they have contracted from their providers. When there is no congestion, the users can send beyond their expected capacity and the extra amount of data is not charged. In this way, the provider must supply enough resources to satisfy the expected capacity from all its customers, and thus its provisioning costs directly relate to the total portion of the expected capacity that has been sold. The main limitations of this scheme are in that traffic profile are not established per-connection, but holds on a long-term basis. This means that in the end, customers pay for a certain access rate. In the NPI context, prices are established and negotiated per-connection and on-demand by considering the current network state.

Wang and Peha give an interesting analysis of state-dependent pricing and its economic implications [341]. After having proposed in earlier works [342], [228] an iterative procedure for pricing in ATM networks, in [341], they refine both congestion ad hoc control mechanisms and dynamic state-dependent pricing techniques with the aim of maximising the total system social welfare.

Very recently, Singh et al. proposed in [306] an extended version of the expected capacity approach called *dynamic capacity contracting*. This pricing scheme is a congestion-sensitive pricing model designed for the Differentiated Service (DS) architecture of the Internet (see Section 2.1.2). The central idea is that, based on congestion monitoring mechanisms, a provider can raise prices and vary short-term contracts (corresponding to the expected-capacity profiles proposed by Clark) depending on the current network state. Preliminary experimental results demonstrate the potential of this approach, even though several open issues still need more focus.

A.1.2 Pricing in Multi-service and Multi-provider Networks

Pioneer work on pricing policies in multiple service class communication networks has been presented by Cocchi et al. in [55] and [56]. The main idea is to have a multi-class network service discipline which uses a FIFO queue that is able to discard packets depending on whether packets have priority flags set. Prices are differentiated on the basis of the required priority: service requiring a lower priority are cheaper. Using simulations, this approach has been demonstrated to set prices in a way that users are more satisfied with the combined cost and performance than when using flat-rate pricing without service differentiation. *Service-class sensitive pricing* is also proposed by Gupta in [116] where Internet traffic is differentiated according to delay and loss requirements.

In [303], Shenker et al. discuss and criticise various pricing aspects considered in previous works, including their own preliminary ideas presented in [302]. Their main critique is for pricing mechanisms aiming at achieving optimal efficiency. Optimality requires indeed to set usage-based charges equal to the marginal costs of usage. But it is argued that this may be infeasible or not appropriate for three main reasons:

- Marginal cost prices may not produce sufficient revenue to fully recover costs.

- Congestion costs are inherently inaccessible to the network (especially in a multi-provider environment) and thus they cannot reliably form the only basis for pricing.

- Besides optimality, there can be other more structural goals (for instance, to achieve incentives to use multi-cast where appropriate) that are incompatible

with the global uniformity of pricing policies in all distinct networks involved
required for optimal pricing schemes.

Then, Shenker and its colleagues propose in [303] a pricing paradigm called the
edge pricing that focuses on architectural issues, rather than on detailed calcula-
tion of marginal congestion cost. The term they propose *"refers to where charges
are accessed rather than their form (e.g., usage-based or not) or their relationship
to congestion"* [303]. The fundamental idea beyond this model is to charge the end
user only by the first network provider (see the *initiator* NPA in this dissertation)
along the selected end-to-end path that might span several domains and use services from other providers. The charge to be paid includes the fees for all distinct
providers involved handling their respective resources.

The inter-domain charging structure proposed in the NPI context is very
similar to this latter approach. The price paid by an EUA to the *initiator* for
the allocation of a specific service demand includes all expenses for crossing all
providers domains along the selected abstract path. The main advantage of this
kind of architecture is that the local pricing process can be performed by every
provider independently on how others decide to estimate prices. Notice that in
the NPI paradigm there are no constraints on which specific negotiation strategy
or pricing policy providers should individually adopt. In addition, this approach
simplifies the billing and accounting procedure from the users perspective. The
principal difference between the original edge pricing scheme and our approach
concerns the way the total price is computed by the *initiator*. In the original edge
pricing model, the *initiator* computes the total price for a service by considering
static pre-existing bilateral agreements with peer providers. This means that the
price is based on the *expected* congestion along the *expected* path and there are
no *"per-flow settlement payments"*. In the NPI context, dynamic peer agreements
allow the computation of prices that do reflect the current criticalness of the
network resources deployed along the on-demand selected abstract path. Although
this requires extra communication and dynamic arrangements between providers,
automated and interoperable software agents have been demonstrated to obtain
good performance (see Chapter 7).

A.1.3   Auction Based Models

As anticipated in Section 2.4.2, auction-based techniques have been proposed in
several works to define pricing mechanisms for network components and/or services.

MacKie-Mason and Varian proposed in [189] and [190] the smart-market
pricing approach. The main goal is to define a mechanism able to control the congestion of network resources. This is based on the idea that users should face prices
that reflect the resource cost that they generate. In the smart-market, every packet
contains a bid, and if the packet is accepted, the user pays a clearing price given by
the highest bid among the packets which are dropped. In this way, the approach

becomes incentive compatible in that the optimal strategy for a self-interested user is to set the bid price in each packet equal to the true evaluation. This work made an important contribution in analysing the problem of pricing the Internet and how to charge for transmitting traffic when the network is congested. Nevertheless, the engineering cost of implementing this approach (i.e., sorting packets by bid-price, per packet and per hop accounting) in real systems is prohibitive and therefore the smart-market is not feasible in current networks (see [306] for more details).

Later approaches, reused some of the smart-market approach principles trying to simplify them and render them more feasible for real networks. In [256] and [207], for instance, Reichl, Foukia et al. proposed a pricing scheme for a multi-provider scenario based on an auction model for an Integrated Services network [27]. The developed scheme, called the *Connection-Holder-is-Preferred-Scheme* (CHiPS), relies upon a flow-based dynamic pricing mechanism that makes use of the RSVP protocol [28]. Despite several preliminary well promising experimental results, there are two main aspects that need more clarification. First of all, how do the providers compute reserve prices so that these effectively reflect the current degree of congestion in the network. Second, it is not clear how the dependencies of running separate and parallel auctions with different providers along the path are handled by the end users. This should be carefully considered when establishing the maximum amount of money that users are willing to offer in every auction process.

Lazar and Semret focus on dynamic pricing mechanisms in Differentiated Service (DS) networks [23]. They deploy bandwidth brokers acting as mediators between consumers (or users) and network operators (or raw bandwidth sellers) (see Section 2.2.2). In order to coordinate all these different players the *progressive second price* auction mechanism is proposed [175], [291]. The main novelty is in defining an auction-based scheme for arbitrary size shares of an infinitely divisible resource. Their game theoretical approach is proven to be feasible in real network environments, for what concerns signalling, billing and a priori information load. In addition, the progressive second price auction is proven to be incentive compatible and efficient. In [292], Semret shows that stability can be achieved even for different services classes and service levels affecting each other, *thus the good news is that dynamic market-driven partitioning of network capacity among services appears sustainable.* However, instabilities at the peering level, or macro level (i.e., between distinct providers), arise even in small networks (three network nodes are used in his simulations) and tight provisioning.

In [78], Fankhauser proposes dynamic negotiations of service level agreements between distinct providers in the DS architecture (see also Section 2.2.2). The pricing strategy adopted by every provider is called the *residual bandwidth* pricing. The main idea is to make use of a pricing function (previously proposed by Wand and Crowcroft in [342]) ensuring that prices increase the more the link resources are used. Costs are assumed to be constant, therefore, to reflect the current resource use Fankhauser proposes to dynamically vary the base prices. In the following, how the intra-domain pricing scheme proposed for provider agents in the NPI framework deploys the criticalness factor of every link deployed in order to dynamically estimate the cost of resources.

Appendix B

Acronyms

The acronyms used in this dissertation are all listed below in alphabetical order.

ACE Availability Criticalness Evaluation
ACL Agent Communication Language
AI Artificial Intelligence
AMI Agent Management Interface
ATM Asynchronous Transfer Mode
BB Bandwidth Broker
BGP Border Gateway Protocol
BI Blocking Island
BIG Blocking Island Graph
BIH Blocking Island Hierarchy
CAC Call Admission Control
CBR Constant Bit Rate
CLA Coalition Leader Agent
CMIP Common Management Information Protocol
CMIS Common Management Information Service
CR Coalition Revenue
COB COalition Based approach
CSP Constraint Satisfaction Problem
DAI Distributed Artificial Intelligence
DCSP Distributed Constraint Satisfaction Problem
DF Directory Facilitator
DS Differentiated Services
DWDM Dense Wavelength Division Multiplexing
EUA End User Agent
FAS Fixed Agreements Solution

FIPA Foundation for Intelligent Physical Agents
IETF Internet Engineering Task Force
IP Internet Protocol
ITU International Telecommunication Union
KQML Knowledge Query and Manipulation Language
LNC Layer Network Coordinator
LNFed Layer Network Federation
MAS Multi-Agent System
MIB Management Information Base
MPLS Multi-Protocol Label Switching
MuSS Multi-provider Set Up
NOC NO Criticalness policy
NPA Network Provider Agent
NPI Network Provider Interoperability
NPI-mas Network Provider Interoperability multi-agent system
NPI-TG Network Provider Interoperabilit Traffic Generator
NRM Network Resource Management
OMG Object Management Group
OSPF Open Shortest Path First
PHB Per Hop Behavior
PNNI Private Network-to-Network Interface
OSI Open Systems Interconnection
QoS Quality of Service
RSVP Resource Reservation Protocol
SDH Synchronous Digital Hierarchy
SLA Service Level Agreement
SMK Shared Management Knowledge
SMTP Simple Mail Transfer Protocol
SNMP Simple Network Management Protocol
SPA Service Provider Agent
TCP Transmission Control Protocol
TDM Time Division Multiplexing
TINA Telecommunications Information Networking Architecture
TMF TeleManagement Forum
TMN Telecommunication Management Network
VBR Variable Bit Rate
VCC Virtual Channel Connection
VPC Virtual Path Connection
VPN Virtual Private Network
WAN Wide Area Network

Bibliography

[1] AgentCities. http://www.agentcities.org.

[2] Dean Allemang and Beat Liver. Functional representation for reusable components. In *Proceedings of the Seventh Workshop on Institutionalizing Software Reuse*, 1995.

[3] E. M. Arkin and E. L. Silverberg. Scheduling jobs with fixed start and end times. *Disc. Appl. Math*, 18:1–8, 1987.

[4] Kenneth J. Arrow. Current developments in the theory of social choice. In *Social Choice and Justice*, chapter 12, pages 162–174. The Belknap Press, 1983.

[5] Kenneth J. Arrow. A difficulty in the concept of social welfare. In *Social Choice and Justice*, chapter 1, pages 1–29. The Belknap Press, 1983.

[6] O. Ashenfelter. How auction work for wine and art. *Journal of Economic Perspectives*, 3(3):23–36, 1989.

[7] M. J. Atallah. *Handbook on Algorithms and Theory of Computation*. CRC Press, 1999.

[8] The ATM Forum. *P-NNI 1.0 Specification*, May 1996.

[9] D. Awduche, A. Chiu, A. Elwalid, I. Widjaja, and X. Xiao. A framework for internet traffic engineering. Internet draft draft-ietf-tewg-framework-02.txt, work in progress, 2000.

[10] D. Awduche, T. Li, and G. Swallow. Igp requirements for traffic engineering with mpls. *Internet Draft, draft-li-mpls-igp-te-00.txt, Internet Engineering Task Force. Work in progress. 12*, 1999.

[11] Baruch Awerbuch, Yi Du, Bilal Khan, and Yuval Shavitt. Routing through networks with hierarchical topology aggregation. Technical Report 98-16, DIMACS, March 5 1998. Mon, 14 Sep 1998 20:00:00 GMT.

[12] Mario Baldi, Silvano Gai, and Gian Pietro Picco. Exploiting code mobility in decentralized and flexible network management. In *Proceedings of the First International Workshop on Mobile Agents*, Berlin, Germany, April 1997.

[13] Mihai Barbuceanu and Wai-Kau Lo. A multi-attribute utility theoretic negotiation architecture for electronic commerce. In *Proceedings of the Fourth International Conference on Autonomous Agents*, pages 239–248, Barcelona, Catalonia, Spain, 2000. ACM Press.

[14] H. Baumgärtel, S. Bussmann, and M. Klosterberg. Combining multi-agent systems and constraint techniques in production logistics. In: *Proc. of the 6th Annual Conf. on AI, Simulation and Planning in High Autonomy Systems*, La Jolla, Ca, USA, pages 361–367, 1996.

[15] Synchronous optical network (SONET) transport systems: Common generic criteria. Technical Report TR-NWT-000253, Issue 2, Bellcore, 1991.

[16] Richard E. Bellman. On a routing problem. *Quarterly of Applied Mathematics*, 16(1):87–90, 1958.

[17] Dimitri Bertsekas and Robert Gallager. *Data Networks*. Prentice Hall, 2nd edition, 1992.

[18] J. Bigham, L. Cuthbert, A. Hayzelden, and Z. Luo. Multi-agent system for network resource management. *Lecture Notes in Computer Science*, 1597:514–526, 1999.

[19] K. Binmore. *Essays on the Foundations of Game Theory*. Blackwell Publishers, Oxford, 1990.

[20] K. Binmore and P. Dasgupta. *The economics of Bargaining*. Blackwell Publishers, Oxford, 1989.

[21] K. Binmore and N. Vulkan. Applying game theory to automated negotiation. In *DIMACS Workshop on economics, Game Theory and the Internet*, Rutgers University, apr 1997.

[22] K. Binmore and N. Vulkan. Applying game theory to automated negotiation. 1997.

[23] S. Blake, D. Black, M. Carlson, E. Davies, and Z. Wang. Weiss: An architecture for differentiated services; rfc, 1998.

[24] S. Bodamer. Charging in Multi-Service Networks. Technical report, Internal Report No. 29, Institute of Communication Networks and Computer Engineering, University of Stuttgart, Stuttgart, Germany., 1998.

[25] Alan H. Bond and Les Gasser. Themes in distributed artificial intelligence. In Alan H. Bond and Les Gasser, editors, *Readings in Distributed Artificial Intelligence*, pages vii–xv. Morgan Kaufmann publishers Inc.: San Mateo, CA, USA, 1989.

[26] Jean-Yves Le Boudec and Tony Przygienda. A Route Pre-Computation Algorithm for Integrated Services Network. Technical Report TR-95/113, LRC-DI, EPFL, Lausanne, Switzerland, 1995.

[27] R. Braden, D. Clark, and S. Shenker. RFC 1633: Integrated services in the Internet architecture: an overview, 1994.

[28] R. Braden, L. Zhang, S. Berson, and S. Herzog, S. andJamin. Resource ReSerVation Protocol (RSVP) – Version 1 Functional Specification. RFC 2205, IETF Network Working Group, Standards track, September 1997.

[29] J. Bradshaw. *An Introduction to Software Agents*. CA: AAAI Press, 1997.

[30] J. Bradshaw. Terraforming cyberspace. *Computer IEEE*, 34(7):48–56, 2001.

[31] P. Bratley, B. L. Fox, and L. E. Schrage. *A guide to simulation*. Springer-Verlag, 1987.

[32] M. Breugst and T. Magedanz. On the Usage of Standard Mobile Agent Platforms in Telecommunication Environments. In S. Trigila et al., editor, *Proceedings of 5th Int. Conference on Intelligence in Services and Networks (IS&N)*, Lecture Notes of Computer Sciences 1430., pages 275–286, Antwerp, Belgium, May 1998. Springer Verlag.

[33] M. Calisti and B. Faltings. A multi-agent paradigm for the Inter-domain Demand Allocation process. *DSOM'99, Tenth IFIP/IEEE International Workshop on Distributed Systems: Operations and Management*, 1999.

[34] M. Calisti, B. Faltings, and Mazziotta S. Market-skilled agents for automating the bandwidth commerce. In *Proceedings of the 3rd IFIP/GI International Conference on Trends towards a Universal Service Market*. Springer-Verlag, September 2000.

[35] M. Calisti, C. Frei, and B. Faltings. A distributed approach for QoS-based multi-domain routing. *AiDIN'99, AAAI-Workshop on Artificial Intelligence for Distributed Information Networking*, 1999.

[36] M. Calisti, S. Willmott, and B. Faltings. Supporting Interworking among Network Providers using a Multi-agent Architecture. *NOC'99, European Conference on Networks and Optical Communications*, 1999.

[37] R. W. Callon. RFC 1195: Use of OSI IS-IS for routing in TCP/IP and dual environments, December 1990. Status: PROPOSED STANDARD.

[38] Kathleen M. Carley and Les Gasser. Computational organization theory. In Gerhard Weiss, editor, *Multiagent Systems: A Modern Approach to Distributed Artificial Intelligence*, chapter 7, pages 299–330. The MIT Press, Cambridge, Massachusetts, 1999.

[39] J. D. Case, M. Fedor, M. L. Schoffstall, and C. Davin. RFC 1098: Simple network management protocol (SNMP), April 1989. Obsoleted by RFC1157. Obsoletes RFC1067. Status: UNKNOWN.

[40] J. D. Case, M. Fedor, M. L. Schoffstall, and C. Davin. RFC 1157: Simple network management protocol (SNMP), May 1990. See also STD0015 . Obsoletes RFC1098 [39]. Status: STANDARD.

[41] R. Cassady. Auction and auctioneering. In Univ. California Press, editor, *Reading*, Oct 1979.

[42] J. Cavanagh. *Frame Relay Applications: Business and Technical Case Studies*. Morgan Kaufman, 1998.

[43] K. Mani Chandy and Leslie Lamport. Distributed snapshots: Determining global states of distributed systems. *TOCS*, 3(1):63–75, February 1985.

[44] Shigang Chen and Klara Nahrstedt. An Overview of Quality of Service Routing for Next-Generation High-Speed Networks: Problems and Solutions. *IEEE Network*, November/December 1998.

[45] Chih-Che Chou and Kang G. Shin. A distributed table-driven route selection scheme for establishing real-time video channels. In *International Conference on Distributed Computing Systems*, pages 52–59, 1995.

[46] Berthe Y. Choueiry and Dean Allemang. Abstraction methods for resource management in a distributed information network. In *Workshop on Artificial Intelligence in Distributed Information Networks, IJCAI'95*, pages 99–102, 1995.

[47] Berthe Y. Choueiry and Toby Walsh, editors. *Abstraction, Reformulation and Approximation: Proceedings of the 4th International Symposium, SARA 2000*, Horseshoe Bay, Texas, USA, July 2000. Lecture Notes in Artificial Intelligence, Vol. 1864, Springer-Verlag.

[48] W. W. Chu, Q. Chen, and R. C. Lee. Cooperative query answering via type abstraction hierarchy. In S. M. Deen, editor, *CKBS-90 — Proceedings of the International Working Conference on Cooperating Knowledge Based Systems*, pages 271–292. Springer-Verlag: Heidelberg, Germany, 1991.

[49] D. Clark. Internet cost allocation and pricing. *In Internet Economics (1997), J. B. L. McKnight, Ed., MIT Press, pp. 215–253.*, 1997.

[50] D. Clark and J. Wroclawski. An approach to sevice allocation in the internet, 1997.

[51] David D. Clark and Wenjia Fang. Explicit allocation of best-effort packet delivery service. *IEEE/ACM Transactions on Networking*, 6(4):362–373, August 1998.

[52] E. H. Clarke. Multipart pricing of public goods. In *Public Choice*, pages 17–33, 11 1971.

[53] E. H. Clarke. Multipart pricing of public goods: An example. In *Public Prices for Public Products*. S. Muskin, 1972.

[54] S. H. Clearwater. *Market Based Control: A paradigm fo Distributed Resource Allocation*. World Scientific, Singapore, 1996.

[55] Ron Cocchi, Deborah Estrin, Scott Shenker, and Lixia Zhang. A study of priority pricing in multiple service class networks. *ACM SIGCOMM 91*, 1991.

[56] Ron Cocchi, Scott Shenker, Deborah Estrin, and Lixia Zhang. Pricing in computer networks: Motivation, formulation, and example. *IEEE Transactions on Networking*, 1(6):614–627, December 1993.

[57] S. E. Conry, K. Kuwabara, V. R. Lesser, and R. A. Meyer. Multistage negotiation for distributed satisfaction. *IEEE Transactions on Systems, Man and Cybernetics*, 21(6):1462–1477, November/December 1991.

[58] S. Corley, M. Tesselaar, J. Cooley, and J. Meinkoehn. The application of intelligent and mobile agents to network and service management. *Lecture Notes in Computer Science*, 1430, 1998.

[59] E. Crawley. A framework for qos-based routing in the internet. *Internet draft, IETF, July 10, 1998. (draft-ietf-qosr-framework-06.txt).*, 1998.

[60] R. Cremonini, K. Marriott, and H. Sndergaard. A general theory of abstraction. In *In Proceedings of the 4th Australian Joint Conference on Artificial Intelligence, pages 121–134, Australia.*, 1990.

[61] D. Curry, M. Menasco, and J. Van Ark. Multi-attribute dyadic choice. Models and tests. *Journal of Marketing Research*, 28(3):259–267, 1991.

[62] Randall Davis and Reid G. Smith. Negotiation as a metaphor for distributed problem solving. *Artificial Intelligence*, 20(1):63–109, 1983.

[63] Martin de Prycker. *Asynchronous Transfer Mode: Solution for Broadband ISDN*. Prentice Hall, 3rd edition, 1995.

[64] Decasper, Dittia, Parulkar, and Plattner. Router plugins: A software architecture for next-generation routers. *IEEETNWKG: IEEE/ACM Transactions on NetworkingIEEE Communications Society, IEEE Computer Society and the ACM with its Special Interest Group on Data Communication (SIGCOMM), ACM Press*, 8, 2000.

[65] Rina Dechter. Constraint networks. In Stuart C. Shapiro, editor, *Encyclopedia of Artificial Intelligence*, pages 276–285. Wiley, 1992. Volume 1, second edition.

[66] R. Diaz-Caldera, J. Serrat-Fernandez, K. Berdekas, and F. Karayannis. An Approach to the Cooperative Management of Multitechnology Networks. *IEEE Communications Magazine*, 37(5):119–125, May 1999.

[67] E. W. Dijkstra. A note on two problems in connection with graphs. *Numerische Mathematik*, 1:269–271, 1959.

[68] W. Donnelly. Managing multiprovider networks. *Computer Communications, Elsevier Science*, 22:1628–1632, 1999.

[69] E. Durfee, J. Lee, and P. J. Gmytrasiewicz. Overeager reciprocal rationality and mixed strategy equilibria. In *Proceedings of the 12th International Workshop on Distributed Artificial Intelligence*, pages 109–129, Hidden Valley, Pennsylvania, May 1993.

[70] Edmund H. Durfee, Victor R. Lesser, and Daniel D. Corkill. Cooperative distributed problem solving. In Avron Barr, Paul R. Cohen, and Edward A. Feigenbaum, editors, *The Handbook of Artificial Intelligence*, volume IV, chapter XVII, pages 83–137. Addison-Wesley, 1989.

[71] eBiz Networks. http://www.bcr.com/ebiznets/.

[72] M. El-Darieby and J. Rolia. Performance modeling of a service provisioning design. In *Lecture Notes in Computer Science, vol, 1890, Springer-Verlag, 2000.* 2000.

[73] Eithan Ephrati and Jeffrey S. Rosenschein. The Clarke tax as a consensus mechanism amoung automated agents. In *Proceedings of the Ninth National Conference on Artificial Intelligence*, pages 173–178, Anaheim, CA, July 1991.

[74] Deborah Estrin, Yakov Rekhter, and Steven Hotz. Scalable inter-domain routing architecture. In *SIGCOMM*, pages 40–52, 1992.

[75] P. Th. Eugster, R. Guerraoui, and J. Sventek. Distributed asynchronous collections: Abstractions for publish/subscribe interaction. In *14th European Conference on Object Oriented Programming (ECOOP 2000), Cannes, France.*, 2000.

[76] Wenjia Fang. *Differentiated Services: Architecture, Mechanisms and an Evaluation.* PhD thesis, Princeton University, 2000.

[77] Wenjia Fang and Larry Peterson. Inter-AS traffic patterns and their implications. Technical Report TR-598-99, Department of Computer Science, Princeton University, March 1999; 20 Pages.

[78] G. Fankhauser. *A Network Architecture Based on Market Principles.* PhD thesis, Swiss Federal Institute of Technology of Zurich (ETH), Switzerland, 2000.

[79] G. Fankhauser and B. Plattner. Diffserv bandwidth brokers as mini-markets. In *Proc. of the International Workshop on Internet Service Quality Economics*, December 1999.

[80] G. Fankhauser, B. Stiller, and B. Plattner. Arrow: A flexible architecture for an accounting and charging infrastructure in the next generation internet. *NET-NOMICS: Economic Research and Electronic Networking*, 1(2), December 1998.

[81] G. Fankhauser, B. Stiller, C. Vgtli, and B. Plattner. Reservation based charging in an integrated services network. In *Proc. of INFORMS Telecommunications, Bocca Raton, FL.*, 1998.

[82] P. Faratin. *Automated Service Negotiation Between Autonomous Computational Agents*. PhD thesis, University of London, Queen Mary College, Department of Electronic Engineering, 2000.

[83] P. Faratin, N. R. Jennings, P. Buckle, and C. Sierra. Automated Negotiation for Provisioning Virtual Private Networks Using FIPA -Compliant Agents. In Jeffrey Bradshaw and Geoff Arnold, editors, *Proceedings of the 5th International Conference on the Practical Application of Intelligent Agents and Multi-Agent Technology (PAAM 2000)*, pages 185–202, Manchester, UK, April 2000. The Practical Application Company Ltd.

[84] P. Faratin, C. Sierra, N. Jennings, and P. Buckle. Designing responsive and deliberative automated negotiators. In *Proc. AAAI Workshop on Negotiation: Settling Conflicts and Identifying Opportunities*, 1999.

[85] P. Faratin, C. Sierra, and N. R. Jennings. Negotiation decision functions for autonomous agents. *Int. Journal of Robotics and Autonomous Systems*, 24(3-4):159–182, 1998.

[86] Peyman Farjami, Carmelita Gorg, and Frank Bell. Advanced service provisioning based on mobile agents. In Ahmed Karmouch and Roger Impley, editors, *First International Workshop on Mobile Agents for Telecommunication Applications (MATA'99)*, pages 259–272, Ottawa, Canada, 1999. World Scientific Publishing Ltd.

[87] Fipa. Agent specification 1997. *Foundation for Intelligent Physical Agents*, October 1997.

[88] Fipa. Network management and provisioning, specification 97, part 7. *Foundation for Intelligent Physical Agents*, 1997.

[89] Fipa. Agent specification 1998. *Foundation for Intelligent Physical Agents*, October 1998.

[90] ATM Forum. http://www.atmforum.com.

[91] ATM FORUM. P-NNI V1.0 - ATM Forum approved specification, af-pnni-0055.000. *ATM FORUM*, 1996.

[92] Foundation for Intelligent Physical Agents. Specifications, 1997. http://www.fipa.org/spec/.

[93] C. Frei. *Abstraction Techniques for Resource Allocation in Communication Networks*. PhD thesis, Artificial Intelligence Laboratory - Swiss Federal Institute of technology of Lausanne, 2000.

[94] C. Frei and B. Faltings. Abstraction and constraint satisfaction techniques for planning bandwidth allocation. In Sahin Albayrak and Francisco J. Garijo, editors,

Proceedings of the 2nd International Workshop on Intelligent Agents for Telecommunication Applications (IATA-98), volume 1437 of *LNAI*, pages 1–16, Berlin, July 4–7 1998. Springer.

[95] C. Frei and B. Faltings. Resource Allocation in Networks Using Abstraction and Constraint Satisfaction Techniques. In *Fifth International Conference on Principles and Practice of Constraint Programming (CP'99)*, pages 204–218, Alexandria, Virginia, October 1999. Springer Verlag.

[96] E. Freuder. Backtrack-free and backtrack-bounded search. In Laveen Kanal and Vipin Kumar, editors, *Search in Artificial Intelligence*. Springer-Verlag, New York, 1988.

[97] Eugene C. Freuder. Direct independence of variables in constraint satisfaction problems. Technical Report 84-15, University of New Hampshire, Department of Computer Science, March 1984.

[98] Eugene C. Freuder and Richard J. Wallace. Suggestion strategies for constraint-based matchmaker agents. In *Principles and Practice of Constraint Programming*, pages 192–204, 1998.

[99] D. Fudenberg and J. Tirole. *Game Theory*. The MIT Press, Cambridge, Massachusetts, 1991.

[100] ITU-T Recommendation G.707. Network node interface for the synchronous digital hierarchy, 1996.

[101] A. Galis, C. Brianza, C. Leone, and C. Salvatori. Towards integrated network management for ATM and SDH networks supporting a global broadband connectivity management service. *Lecture Notes in Computer Science*, 1238, 1997.

[102] A. Galis and D. Griffin. A comparison of approaches to multi domain connection management.

[103] Michael R. Garey and David S. Johnson. *Computers and Intractibility, A Guide to the Theory of* NP-*Completeness*. W. H. Freeman and Company, New York, 1979.

[104] Alan Garvey, Keith Decker, and Victor Lesser. A negotiation-based interface between a real-time scheduler and a decision-maker. Technical Report UM-CS-1994-008, University of Massachusetts, Amherst, Computer Science, February, 1994.

[105] Michael R. Genesereth, Matthew L. Ginsberg, and Jeffrey S. Rosenschein. Cooperation without communication. In *Proceedings AAAI-86*, pages 561–567, 1986.

[106] M. Gerla, H. Chan, and J. de Marca. Routing, flow control and fairness in computer networks. In *Proc. IEEE Internatl. Conf. Comm.*, pages 1272–1275, 1984.

[107] R. Gibbsons. *Game Theory for Applied Economists*. Princeton University Press., 1992.

[108] M. A. Gibney and N. R. Jennings. Dynamic resource allocation by market-based routing in telecommunications networks. *Lecture Notes in Computer Science*, 1437, 1998.

[109] F. Giunchiglia, T. Walsh, and A. of. A theory of abstraction. *Articial Intelligence*, 56(2–3):323–390, 1992., 1992.

[110] I. J. Good. Twenty-seven principles of rationality. In V. P. Godambe and D. A. Sprott, editors, *Foundations of statistical inference*, pages 108–141. Holt Rinehart Wilson, Toronto, 1971.

[111] R. Govindan, C. Alaettinoglu, G. Eddy, D. Kessens, S. Kumar, and W. Lee. An architecture for stable, analyzable internet routing. *IEEE Network Magazine*, January/February 1999.

[112] S. Green, L. Hurst, B. Nangle, P. Cunningham, F. Somers, and R. Evans. Intelligent agents group report, 1997.

[113] D. Griffin, G. Pavlou, and T. Tin. Implementing TMN-like management services in a TINA compliant architecture: A case study on resource configuration management. *Lecture Notes in Computer Science*, 1238:263, 1997.

[114] T. Groves. Optimal allocation of public goods: A solution to the free rider problem. *Econometrica*, 45,783–809., 1977.

[115] R. Guerraoui. Fault-tolerance by replication in distributed systems - a tutorial. In *International Conference on Reliable Software Technologies (Invited Paper)*. Springer and Verlag (LNCS), 1996.

[116] Alok Gupta, Dale O. Stahl, and Andrew B. Whinston. Priority pricing of integrated services networks. In *Internet Econnomics*, 1996.

[117] T. Hamada, H. Kamata, and S. Hogg. An Overview of the TINA MAnagement Architecture. *Journal of Network and System Management*, 5(4):411–435, 1997.

[118] Y. Hamadi, C. Bessière, and J. Quinqueton. Backtracking in distributed constraint networks. In Henri Prade, editor, *Proceedings of the 13th European Conference on Artificial Intelligence (ECAI-98)*, pages 219–223, Chichester, August23–28 1998. John Wiley and Sons.

[119] Walter C. Hamscher. Modeling digital circuits for troubleshooting. *Artificial Intelligence*, 51(1–3):223–271, 1991.

[120] B. Hayes-Roth, F. Hayes-Roth, S. Rosenschein, and S. Cammarata. Modelling planning as an incremental. In *Proc. of the International Joint Conference on Artificial Intelligence, 1979.*, 1979.

[121] B. Hayes-Roth, M. Hewett, R. Washington, R. Hewett, and A. Seiver. Distributing intelligence within an individual. Technical Report CS-TR-88-1229, Stanford University, Department of Computer Science, November 1988.

[122] A. L. G. Hayzelden and J. Bigham. Heterogeneous Multi-Agent Architecture for ATM Virtual Path Network Resource Configuration. In Sahin Albayrak and Francisco J. Garijo, editors, *Proceedings of the 2nd International Workshop on Intelligent Agents for Telecommunication Applications (IATA-98)*, volume 1437, pages 45–59. Springer-Verlag: Heidelberg, Germany, 4–7 1998.

[123] Alex L. G. Hayzelden and John Bigham. Agent Technology in Communications Systems: An Overview. *Knowledge Engineering Review*, 1999.

[124] Alex L. G. Hayzelden and John Bigham, editors. *Software Agents for Future Communication Systems: Agent Based Digital Communication*. Springer-Verlag, Berlin Germany, 1999.

[125] Hayzelden, A. L. G. and Bigham, J. *Software Agents for Future Communication Systems*. Springer Verlag, April 1999.

[126] C. Hedrick. Routing information protocol. RFC 1058, IETF Network Working Group, June 1988.

[127] C. L. Hedricks. An Introduction to IGRP. Technical report, Laboratory for Computer Science Research, Rutgers University., August 1991.

[128] H. Hegering, S. Abeck, and B. Neumair. *Integrated Management of Networked Systems*. Morgan Kaufmann Publishers, 1999.

[129] J. Heinanen, F. Baker, W. Weiss, and J. Wroclawski. Assured forwarding phb group, 1999.

[130] R. Heiner. The origin of predictable behavior. *American Economic Review*, 73: 560–595., 1983.

[131] A. Hopson and R. Janson. Deployment scenarios for interworking. Engineering Notes TP_AJH.001_0.10_94, TINA-C, 1995.

[132] Qianbo Huai and Tuomas Sandholm. Nomad: Mobile Agent System for an Internet-Based Auction House. *Internet Computing*, 4(2):80–86, / 2000.

[133] Pu Huang and Katia Sycara. A computational model for online agent negotiation. In *Submitted to IJCAI*, 2000.

[134] Michael N. Huhns and David M. Bridgeland. Multiagent truth maintenance. *IEEE Transactions on Systems, Man, and Cybernetics*, 21(6):1437–1445, 1991.

[135] IATA. *Intelligent Agent for Telecommunications Applications*. 1996.

[136] T. Ibaraki and N. Katoh. *Resource Allocation Problems: Algorithmic Approaches*. MIT Press, Cambridge, MA, 1988.

[137] Internet2. http://www.internet2.edu/.

[138] ISO/IEC International Standard, IS9595. Common management information service definition, 1995.

[139] ISO/IEC International Standard, IS9596. Common management information protocol specification, 1996.

[140] ITU-T, editor. *ITU-T Recommendation Principles for a Telecommunications management network, M3010*. Genf, 1996.

[141] Reccomendation E.800 ITU-T, editor. *Telephone Network and ISDN. Quality of Service, network management and traffic engineering. Terms and definitions related to Quality of service ad Network Performance including dependability*. August, 1994.

[142] Reccomendation I.350 ITU-T, editor. *General aspects of Quality of Service and Network performance in Digital networks, including ISDN*. 1993.

[143] Recommendation M3020 ITU-T, editor. *TMN interface specification methodology*. 1995.

[144] Recommendation M3320 ITU-T, editor. *Management requirements frameowrk for the TMN-X interface*. 1997.

[145] J. Park J. Won-Ki Hong. Implementation and performance of a tmn mk system. In *Seventh IFIP/IEEE International Workshop on Distributed Systems: Operations and Management, L'Aquila (Italy)*, 1996.

[146] V. Jacobson, K. Nichols, and K. Poduri. An expedited forwarding phb, 1999.

[147] Raj Jain. The art of computer systems performance analysis: Techniques for experimental design, measurement, simulation and modeling (book review). *SIGMETRICS Performance Evaluation Review*, 19(2):5–11, 1991.

[148] N. Jennings and M. Wooldrige. *Agent Technology Foundations, Applications, and Markets*. Springer/UNICOM, February 1998.

[149] N. R. Jennings. Building complex software systems. *Communications of the ACM*, 44(4):35–41, 2001.

[150] N. R. Jennings, P. Faratin, A. R. Lomuscio, S. Parsons, C. Sierra, and M. Wooldridge. Automated Negotiation: Prospects, Methods and Challenges. *In. Journal of Group Decision and Negotiation*, 10(2):199–215, 2001.

[151] N. R. Jennings, T. J. Norman, and P. Faratin. ADEPT: An agent-based approach to business process management. *ACM SIGMOD Record*, 27(4):32–39, 1998.

[152] L. JyiShane and K. Sycara. Emergent constraint satisfaction through multi-agent coordinated interaction. 1993.

[153] J. Kahan and A. Rapoport. *Theories of Coalition Formation*. 1984.

[154] M. Kahani and H. Beadle. Decentralised approaches for network management. *Computer Communications Review, ACM SIGCOMM, Vol. 27 N.3 July*, 1997.

[155] F. Karayannis, K. Berdekas, R. Diaz, and J. Serrat. A telecommunication operators interdomain interface enabling multi-domain. *Multi-technology Network Management, Interoperable Communication Networks Journal, Vol. 2, No. 1, pp. 1–10, Baltzer Science Publishers, March 1999.*, 1999.

[156] Kazovsky, L. and Benedetto, S. and Willner, A. *Optical Fibre Communication Systems*. Artech House, Boston, 1906.

[157] C. Knoblock. Automatically generating abstractions for problem solving. Technical report cmu-cs-91-120, school of computer science, carnegie mellon university, 1991., 1991.

[158] Donald E. Knuth. *The Art of Computer Programming, Volume 1, Fundamental Algorithms*. Addison-Wesley, Reading, MA, USA, 1973.

[159] T. C. Koopmans. Uses of price. In T. C. Koopmans, editor, *Scientific Papers of Tjalling C. Koopmans*, pages 243–257. Springer-Verlag, 1970.

[160] R. Korf. Planning as search: A qualitative approach. *Artificial Intelligence, 33:65–88, 1987.*, 1987.

[161] R. Kowalczyk and V. Bui. On Constraint-Based Reasoning in e-Negotiation Agents. In F. Dignum and U. Cortes, editors, *Agent-Mediated Electronic Commerce III: Current Issues in Agent Based Electronic Commerce Systems*, page 31 ff. Springer Verlag, 2001.

[162] S. Kraus and J. Wilkenfeld. Multiagent negotiation under time constraints. Technical Report CS 2649, Institute for Advanced Computer Studies, University of Maryland, 1993.

[163] S. Kraus, J. Wilkenfeld, and G. Zlotkin. Multi Agent Negotiation Under Time Constraints. *Artificial Intelligence*, 75(2):297–345, 1995.

[164] Sarit Kraus, Katia Sycara, and Amir Evenchik. Reaching agreements through argumentation: A logical model and implementation. *Artificial Intelligence*, 104(1–2):1–69, 1998.

[165] Sarit Kraus, Jonathan Wilkenfeld, and Gilad Zlotkin. Multiagent negotiation under time constraints. *Artificial Intelligence*, 75(2):297–345, 1995.

[166] David M. Kreps. *A Course in Microeconomic Theory*. Princeton University Press, Princeton, NJ, USA, 1990.

[167] G. P. Kumar and P. Venkataram. Artificial Intelligence Approaches to Network Management: Recent Advances and a Survey. *Computer Communications*, 20:1313–1322, 1997.

[168] M. Kumar and S. I. Feldman. Internet auctions. Technical report, IBM, 1998.

[169] Vipin Kumar. Algorithms for Constraint Satisfaction Problems: A Survey. *AI Magazine*, 13(1):32–44, 1992.

[170] K. Kuwabara, T. Ishida, Y. Nishibe, and T. Suda. An Equilibratory Market Based Approach for Distributed Resource Allocation and its Applications to Communication Network Control. In S. H. Clearwater, editor, *Market-Based Control: A Paradigm for Distributed Resource Allocation.*, pages 53–73. World Scientific, 1996.

[171] K. Larson and T. Sandholm. Computationally limited agents in auctions. In *Autonomous Agents 2001 Workshop on Agent-Based Approaches to B2B, Montreal, Canada.*, 2001.

[172] K. Larson and T. Sandholm. Costly valuation computation in auctions. In *Theoretical Aspects of Rationality and Knowledge (TARK VIII), Siena, Italy.*, July 2001.

[173] K. Larson and T. Sandholm. Deliberation in Equilibrium: Bargaining in Computationally Complex Problems. *Artificial Intelligence , accepted, to appear.*, 2001.

[174] B. Lassri, H. Laasri, S. Lander, and S. Lesser. A Generic Model for Intelligent Negotiating Agents. *International Journal of Intelligent and Cooperative Information Systems*, 1(2):291–317, 1992.

[175] A. Lazar and N. Semret. Design and analysis of the progressive second price auction for network bandwidth sharing. *Special issue on Network Economics, 1999. Available as Tech. Rep. CU/CTR/TR 497-98-21.*, 1999.

[176] Ho Guen Lee. Do electronic marketplaces lower the price of goods? *Communications of the ACM*, 41(1):73–80, January 1998.

[177] L. Chi-Hang Lee. *Negotiation Strategies and their Effect in a Model of Multi-Agent Negotiation*. PhD thesis, Department of Computer Science, University of ESSEX, 1996.

[178] W. L. Lee, M. G. Hluchyj, and P. A. Humblet. Routing subject to Quality of Service Constraints. *IEEE Network*, July/August 1995.

[179] V. R. Lesser. Reflections on the nature of multi-agent coordination and its implications for an agent architecture. Technical Report UM-CS-1998-010, University of Massachusetts, Amherst, Computer Science, July, 1998.

[180] A. Y. Levy. Creating abstractions using relevance reasoning. In *Workshop on Theory Reformulation and Abstraction, 1994.*, 1994.

[181] D. Lewis, V. Wade, and R. Bracht. The development of integrated inter and intra domain management services. In *Proceedings of the Sixth IFIP/IEEE International Symposium on Integrated Network Management, IM'99*, 1999.

[182] T. Li and Y. Rekhter. A provider architecture for differentiated services and traffic engineering (paste), October 1998. available at ftp://ftp.isi.edu/in-notes/rfc2430.txt.

[183] J. Liu and K. Sycara. Exploiting problem structure for distributed constraint optimization. In *Proceedings of the First International Conference on Multi-Agent Systems (ICMAS-95)*, pages 246–253, San Francisco, CA, June 1995.

[184] A. Lomuscio, M. Wooldridge, and N. Jennings. A classification scheme for negotiation in eletronic commerce. In F. Dignum and C. Sierra, editors, *Agent-Mediated Electronic Commerce: A European Perspective*, pages 19–33. Springer Verlag, 2001.

[185] K. Lougheed and Y. Rekhter. Border Gateway Protocol. Technical Report 1163, 1990.

[186] N. Luhman. *Trust.* Chicester: John Wiley and Sons., 1979.

[187] Q. Y. Luo, P. G. Hendry, and J. T. Buchanan. Comparison of different approaches for solving distributed constraint satisfaction problems. pages 150–159.

[188] J. MacKie-Mason. A smart market for resource reservation in a multiple quality of service information network. technical report, university of michigan, u.s.a., september 1997. Technical report, university of michigan, u.s.a., september,, 1997.

[189] J. MacKie-Mason and H. Varian. Pricing the internet, public access to the internet. *Public Access to the Internet*, 1993.

[190] J. MacKie-Mason and H. Varian. Pricing congestable network resources. *IEEE Journal on Selected Areas in Communications*, 13:1141–1149, 1995.

[191] Alan Mackworth. Constraint satisfaction. In Stuart C. Shapiro, editor, *Encyclopedia of Artificial Intelligence*, pages 285–293. Wiley, 1992. Volume 1, second edition.

[192] Alan K. Mackworth. Consistency in Networks of Relations. *Artificial Intelligence*, 8:99–118, 1977.

[193] Pattie Maes. Artificial life meets entertainment: Lifelike autonomous agents. *Communications of the ACM*, 38(11):108–114, November 1995.

[194] T. Magendaz and R. Popescu-Zeletin. Intelligent networks-basic technology, styandards and evolution. *Thomson Computer Press, London*, April 1996.

[195] J. Makris and B. Not. Bandwidth brokers, not exactly nasdaq. *Data Communications Magazine*, 1999.

[196] J. Martin-Flatin, S. Znaty, and J. Hubaux. A survey of distributed network and systems management paradigms. Technical report, 1998.

[197] S. Martin-Flatin, J. P.and Znaty. Two Taxonomies of Distributed Network Management Paradigms. In S. Erfani and P. Ray, editors, *Emerging Trends and Challenges in Network Management*. Plenum Press, New York, NY, USA, March 2000.

[198] A. Mas-Collel, M. Whinston, and Jerry Green. *Microeconomic Theory.* 1995.

[199] C. Mason and R. Johnson. Datms: A framework for ditributed assumption based reasoning. In L. Gasser and M. Huhns, editors, *Distributed Artificial Intelligence*, pages 293–318. Morgan Kaufmann, 1989.

[200] N. Mattos and A. Dengel. The role of abstraction concepts and their built-in reasoning in document representation and structuring. In *Proc. of the International Symposium on Artificial Intelligence, pages 136–142. LIMUSA, Balderas, Mexico.*, 1991.

[201] M. S. Miller, D. Krieger, and N. Hardy. An Automated Auction in ATM Network Bandwidth. In S. H. Clearwater, editor, *Market-Based Control: A Paradigm for Distributed Resource Allocation.*, pages 96–125. World Scientific, 1996.

[202] D. L. Mills. RFC 904: Exterior Gateway Protocol formal specification, April 1984. Updates RFC0827, RFC0888. Status: HISTORIC.

[203] T. Mota, P. Hellemans, and T. Tiropanis. Tina as a virtual market place for telecommunication and information services: The vital experiment. In *Proceedings of TINA '99, April,*, 1999.

[204] Alexandros Moukas, Carles Sierra, and Fredrik Ygge, editors. *Agent mediated electronic commerce II: towards next-generation agent-based electronic commerce systems.*, volume 1788 of *Lecture Notes in Computer Science and Lecture Notes in Artificial Intelligence*, New York, NY, USA, 2000. Springer-Verlag Inc.

[205] J. Moy. The OSPF specification. RFC 1131, IETF, October 1989.

[206] Igor Mozetic. Hierarchical model-based diagnosis. *International Journal of Man-Machine Studies*, 35(3):329–362, 1991.

[207] B. Stiller N. Foukia, D. Billard. P. Reichl. User behavior for a pricing scheme in a multi-provider scenario. In *Proc. of the International Workshop on Internet Service Quality Economics*, December 1999.

[208] M. Nakamura, M. Sato, and T. Hamada. A Pricing and Accounting Architecture for QoS Guaranteed Services on a Multi-Domain Network. In *Proceedings of GLOBECOM'99, Global Telecommunication Conference*, pages 1984–1988, 1999.

[209] J. F. Nash. The bargaining problem. *Econometrica*, 18:155–162, 1950.

[210] A. Newell. Limitations of the current stock of ideas about problem-solving, 1965.

[211] A. Newell. The Knowledge Level. *Artificial Intelligence*, 18:87–127, 1982.

[212] K. Nichols, V. Jacobson, and L. Zhang. Rfc 2638: A two-bit differentiated services architecture for the internet, 1999.

[213] Y. Nishibe, K. Kuwabara, T. Ishida, and M. Yokoo. Speed-up of distributed constraint satisfaction and its application to communication network path assignments, 1994.

[214] NMF. Smart tmn telecom operations map, April 1998. Issue 1.

[215] P. Noriega, C. Sierra, M. Giordano, R. de M'antaras, F. Mart'in, and J. Rodr'iguez. The fish market metaphor. Research report iiia-rr-96-17., 1996.

[216] Pablo Noriega and Carles Sierra, editors. *Agent mediated electronic commerce: First International Workshop on Agent-Mediated Electronic Trading, AMET-98: Minneapolis, MN, USA, May 10th, 1998: selected papers*, volume 1571 of *Lecture Notes in Computer Science and Lecture Notes in Artificial Intelligence*, New York, NY, USA, 1999. Springer-Verlag Inc.

[217] H. S. Nwana, L. C. Lee, and N. R. Jennings. Coordination in software agent systems. *The British Telecom Technical Journal*, 14(4):79–88, 1996.

[218] A. Odlyzko. Paris metro pricing: The minimalist differentiated services solution. 1999.

[219] G. M. P. O'Hare and N. R. Jennings. Foundations of distributed artificial intelligence. In *Foundations of Distributed Artificial Intelligence*. John Wiley and Sons, 1996.

[220] OMG. http://www.omg.org, 1989.

[221] M. Osborne and A. Rubinstein. *Bargaining and Markets*. New York: Academic Press., 1990.

[222] P. Cohen. *Empirical Methods for Artificial Intelligence*. MIT Press: Cambridge, MA., 1995.

[223] Project P603. Quality of Service: Measurement Method Selection. Technical Report Deliverable 1, Volume 1, Eurescom, 1997.

[224] D. Parkes and L. Ungar. Iterative combinatorial auctions: Theory and practice. In *Proc. 17th National Conference on Artificial Intelligence (AAAI-00), 74–81.*, 2000.

[225] S. D. Parsons and N. R. Jennings. Negotiation through argumentation-A preliminary report. In *Proc. Second Int. Conf. on Multi-Agent Systems*, pages 267–274, Kyoto, Japan, 1996.

[226] M. Paul. Auctions and bidding: A primer. 1989.

[227] G. Pavlou and D. Griffin. Realizing tmn-like management services in tina. *Journal of Network and System Management (JNSM), Special Issue on TINA, Vol. 5, No. 4, pp. 437–457, Plenum Publishing, 1997.*, 1997.

[228] J. Peha. Dynamic pricing as congestion control in atm networks. In *Proceedings of IEEE GLOBECOM'97, pages 1367–1372*, 1997.

[229] J. Peha and S. Tewari. The results of competition between integrated-services telecommunication carriers. *Information Economics and policy, Special Issue on Multimedia*, 10(1):127–55, March 1998.

[230] Radia Perlman. *Interconnections: Bridges and Routers*. Addison Wesley, 1992.

[231] Radia J. Perlman, Ross W. Callon, and I. Michael C. Shand. Routing architecture. *Digital Technical Journal of Digital Equipment Corporation*, 5(1):62–69, Winter 1993.

[232] David A. Plaisted. Theorem proving with abstraction. *Artificial Intelligence*, 16(1):47–108, March 1981.

[233] S. Poslad and M. Calisti. Towards improved trust and security in fipa agent platforms. In *Autonomous Agents 2000 Workshop on Deception, Fraud and Trust in Agent Societies. June 2000, Barcelona, Spain*, 2000.

[234] S. Poslad, P. Charlton, and M. Calisti. Protecting what your agent is doing. *AgentLink News, Issue 7*, 2001.

[235] S. Poslad, J. Pitt, R. Mamdani, A. Hadingham, and P. Buckle. Agent-Oriented Middleware for Integrating Customer Network Services. In A. Hayzelden and J. Bughma, editors, *Software Agents for Future Communciations Systems*. Springer Verlag, 1999.

[236] Jon Postel, Editor. Transmission Control Protocol — DARPA Internet Program Protocol Specification. Technical Report rfc793, 1981.

[237] P. Pradhan. Pricing in integrated service networks. Technical report.

[238] RACE Project PREPARE. Ibc vpn services, 1993.

[239] P. Princeton and P. Rubinstein. Perfect equilibrium in a bargaining model. *Econometrica* 50(1): 97–109., 1982.

[240] O. Prnjay. Integrity methodology for interoperability environments. *IEEE Communications Magazine*, 37(5):126–132, 1999.

[241] MISA Project. http://www.misa.ch.

[242] The RSVP Project. http://www.isi.edu/div7/rsvp/.

[243] P. Prozeller. TINA and the Software Infrastracture of the Telecom Network of the Future. *Journal of Network and System Management*, 5(4):393–409, 1997.

[244] D. Pruitt. *Negotiation Behavior*. Academic Press, New York, 1981.

[245] R. Braden and L. Zhang and S. Berson and S. Herzog and S. Jamin. *Resource ReSerVation Protocol (RSVP) - Version 1 Functional Specification*. Internet Draft, August 1996.

[246] H. Raiffa. The art and science of negotiation. *MA: Harvard Univ. Press, 1982.*, 1982.

[247] E. Rasmusen. Games and information. *Oxford: Basil Blackwell.*, 1989.

[248] Danny Raz and Yuval Shavitt. Optimal Partition of QoS requirements with Discrete Cost Functions. In *Proceedings of IEEE INFOCOM2000, Nineteenth annual joint conference of the IEEE computer and communication societies*, pages 220–232, July 1999.

[249] I. Recommendation. Recommendation x.700. management framework for open systems interconnection. 1992.

[250] I. Recommendation. Itu-t recommendation i.320. broadband aspects of isdn, 1993.

[251] I. Recommendation. Itu-t: Recommendation i.321. atm bisdn protocol reference model and its application, 1993.

[252] I. Recommendation. Itu-t. recommendation x.723: Generic management information. 1993.

[253] I. Recommendation. Lower layer protocols for the q3 and x interfaces, 1997.

[254] I. Recommendation. Upper layer protocol profiles for the q3 and x interfaces, 1997.

[255] REFORM. http://www.infowin.org/acts/analysis/products/thematic/atm/ch4/reform.html.

[256] P. Reichl and G. Fankhauser. Stiller: Auction models for multiprovider internet connections; messung. In *Messung, Modellierung und Bewertung von Rechensystemen (MMB'99), Trier, Germany, September 21–22,*, 1999.

[257] P. Reichl, S. Leinen, and B. Stiller. A practical review of pricing and cost recovery for internet services. In *Proc. 2nd Internet Economics Workshop Berlin (IEW'99), Berlin, Germany, May,*, 1999.

[258] Y. Rekheter, B. Davie, D. Katz, E. Rosen, and G. Swallow. Cisco systems tag switching architecture overview. Internet rfc 2105., 1997.

[259] Y. Rekhter. Inter-domain routing: Routing: Egp, bgp, and idrp. In Martha Steenstrup, editor, *Routing in Communications Networks*, pages 99–133. Prentice Hall, 1995.

[260] Y. Rekhter. Li: A border gateway protocol, 1995.

[261] Y. Retkher and R. Li. A border gateway protocol 4. Rfc 1771, ibm corporation, cisco systems, 1995.

[262] J. Roberts, U. Mocci, and J. Virtamo. Broadband network traffic: Performance evaluation and design of broadband multiservice networks. *Lecture Notes in Computer Science*, 1155:xxii + 584, 1996.

[263] J. Rodriguez-Aguilar, F. Martin, P. Noriega, P. Garcia, and C. Sierra. Towards a tesbed for trading agents in electronic auction markets. *AI Communications* 11 5–19., 1998.

[264] E. Rosen, A. Viswanathan, and R. Callon. Multiprotocol label switching architecture. *Internet Draft*, 1999.

[265] J. S. Rosenschein and M. Genesereth. Deals among rational agents. In *Proceedings of the Ninth International Conference on Artificial Intelligence*, pages 91–99, Los Angeles, CA, July 1985.

[266] Rosenschein, J. S. and Zlotkin, G. *Rules of Encounter*. MIT Press: Cambridge, MA., 1994.

[267] M. H. Rothkopf and R. M. Harstad. Two models of bid-taker cheating in vickrey auctions. *Journal of Business*, April 1995, 68(2), pp. 257–267., 1995.

[268] M. H. Rothkopf, T. J. Teisberg, and E. P. Kahn. Why are vickrey auctions rare. *A Study in the Economics of Risk, Oxford University Press. Why are Vickrey auctions rare?, Journal of Political Economy 98: 94–109.*, 1990.

[269] A. Rubinstein. *Modeling Bounded Rationality*. MIT Press, 1998., 1998.

[270] Stuart J. Russell and Peter Norvig. *Artificial Intelligence. A Modern Approach*. Prentice-Hall, Englewood Cliffs, 1995.

[271] S. Albayrak and F. J. Garijo, editor. *Intelligent Agent for Telecommunications Applications. Proceedings of the Second International Workshop, IATA'98, Paris, France*. Springer-Verlag Inc., 1998.

[272] S. Albayrak and F. J. Garijo, editor. *Intelligent Agent for Telecommunications Applications. Proceedings of the Third International Workshop, IATA'99, Stockholm, Sweden*. Springer-Verlag Inc., 1999.

[273] Hiroshi Saito and Kohei Shiomoto. Dynamic call admission control in ATM networks. *IEEE Journal of Selected Areas in Communications*, 9(7):982–989, 1991.

[274] T. Sandholm. Issues in computational vickrey auctions. *International Journal of Electronic Commerce, 1999.*, 1999.

[275] Tuomas Sandholm. Automated negotiation. *Communications of the ACM*, 42(3):84–85, March 1999.

[276] Tuomas Sandholm. Agents in electronic commerce: Component technologies for automated negotiation and coalition formation. *Autonomous Agents and Multi-Agent Systems*, 3(1):73–96, March 2000.

[277] Tuomas Sandholm. eMediator: a next generation electronic commerce server. In Carles Sierra, Gini Maria, and Jeffrey S. Rosenschein, editors, *Proceedings of the 4th International Conference on Autonomous Agents (AGENTS-00)*, pages 341–348, NY, June 3–7 2000. ACM Press.

[278] Tuomas Sandholm and Subhash Suri. Improved algorithms for optimal winner determination in combinatorial auctions and generalizations. In *AAAI/IAAI*, pages 90–97, 2000.

[279] Tuomas Sandholm, Subhash Suri, Andrew Gilpin, and David Levine. Winner determination in combinatorial auction generalizations. In *Autonomous Agents 2001 Workshop on Agent-Based Approaches to B2B, Montreal, Canada.*, 2001.

[280] Tuomas W. Sandholm. Limitations of the Vickrey auction in computational multiagent systems. In Victor Lesser, editor, *Proceedings of the First International Conference on Multi–Agent Systems*. MIT Press, 1995.

[281] Tuomas W. Sandholm. *Negotiation Among Self-Interested Computationally Limited Agents*. PhD thesis, University of Massachusetts Amherst, 1996.

[282] Tuomas W. Sandholm. Distributed rational decision making. In MIT Press, editor, *Multiagent Systems: A Modern Introduction to Distributed Artificial Intelligence*, pages 201–258. G. Weiss, 1999.

[283] Tuomas W. Sandholm and Victor R. Lesser. Coalition formation amoung bounded rational agents. In Chris S. Mellish, editor, *Proceedings of the Fourteenth International Joint Conference on Artificial Intelligence*, pages 662–669, Montréal, Canada, August 1995.

[284] Tuomas W. Sandholm and Victor R. Lesser. Advantages of a leveled commitment contracting protocol. In *Proceedings of the Thirteenth National Conference on Artificial Intelligence and the Eighth Innovative Applications of Artificial Intelligence Conference*, pages 126–133, Menlo Park, August 4–8 1996. AAAI Press / MIT Press.

[285] Tuomas W. Sandholm and Victor R. Lesser. Coalitions among computationally bounded agents. *Artificial Intelligence*, 94(1–2):99–137, 1997.

[286] Arvind Sathi and Mark S. Fox. Constraint-directed negotiation of resource reallocations. In Michael N. Huhns and Les Gasser, editors, *Distributed Artificial Intelligence*, volume 2 of *Research Notes in Artificial Intelligence*. Pitman, 1989.

[287] O. Schelen. *Quality of Service Agents in the Internet*. PhD thesis, Dept of Computer Science and Electrical Engineering Luleå University of Technology, 1998.

[288] J.M. Schneider and W. Donnelly. End-to-End Communications Management Using TMN X Interfaces. *Journal of Network and System Management*, 3(1):85–110, 1995.

[289] J. Searle. *Speech Act Theory*. 1969.

[290] N. Semret, M. for, N. Sharing, and P. thesis. *Columbia Univ.* PhD thesis, 1999.

[291] N. Semret, R. R.-F. Liao, A. T. Campbell, and A. A. Lazar. Market pricing of differentiated internet services. In *Proc. of the 7th International Workshop on Quality of Service (IEEE/IFIP IWQOS'99)*, June 1999.

[292] N. Semret, R. R.-F. Liao, A. T. Campbell, and A. A. Lazar. Peering and Provisioning of Differentiated Internet Services. In *INFOCOM - Nineteenth Annual Joint Conference Of The IEEE Computer And Communications Societies*, Tel Aviv, Israel, March 2000. IEEE.

[293] A. Sen. Social choice theory: A re-examination. *Econometrica*, 45:348–384, 1977.

[294] A. Sen. *Social choice theory*. 1986.

[295] Cisco's 12000 series. http://www.cisco.com/warp/public/733/12000.

[296] Juniper's M40 Series, http://www.jumiper.net/products/default.html.

[297] Differentiated Services. http://www.ietf.org/html.charters/diffserv-charter.html.

[298] Integrated Services. http://www.ietf.org/html.charters/intserv-charter.html.

[299] P. Sevcik. ebiz networks: Foundation services for commerce, 2001.

[300] Sexton, M. and Reid, A. *Broadband Networking: ATM, SDH, and SONET*. Artech House, Boston/London, 1997.

[301] Onn Shehory and Sarit Kraus. Task allocation via coalition formation amoung autonomous agents. In Chris S. Mellish, editor, *Proceedings of the Fourteenth International Joint Conference on Artificial Intelligence*, pages 655–661, Montréal, Canada, August 1995.

[302] S. Shenker. Service models and pricing policies for an integrated services internet. In *Public Access to the Internet, MIT Press, Harvard, Cambridge MA, 1995*. 1995.

[303] S. Shenker, D. Clark, D. Estrin, and S. Herzog. Pricing in computer networks: Reshaping the research agenda. *ACM Computer Communications Review*, 26(2):19–43, April 1996.

[304] C. Sierra, N. R. Jennings, P. Noriega, and S. Parsons. A framework for argumentation-based negotiation. *Lecture Notes in Computer Science*, 1365, 1998.

[305] Herbert A. Simon. *Models of Bounded Rationality, Volume 2*. The MIT Press, Cambridge, Massachusetts, 1982.

[306] R. Singh, M. Yuksel, S. Kalyanaraman, and T. Ravichandran. A comparative evaluation of internet pricing models: Smart markets and dynamic capacity contracting.

[307] J. Smith and D. Smith. Database abstractions : aggregation and generalization. *ACM TODS, 2, 2, 1977*, 1977.

[308] R. G. Smith. The contract net protocol: High-level communication and control in a distributed problem solver. In *Proceedings of the 1st International Conference on Distributed Computing Systems*, pages 186–192, Washington, DC, 1979. IEEE Computer Society.

[309] Y.-P. So and E. H. Durfee. A distributed problem solving infrastructure for computer network management. *International Journal of Intelligent and Cooperative Information Systems*, 1(2):363–392, 1992.

[310] Gadi Solotorevsky and Ehud Gudes. Algorithms for solving distributed constraint satisfaction problems (DCSPs). In B. Drabble, editor, *Proceedings of the 3rd International Conference on Artificial Intelligence Planning Systems (AIPS-96)*, pages 191–198. AAAI Press, 1996.

[311] F. Somers. HYBRID: Intelligent Agents for Distributed ATM Network Management. In *Working Notes of the Workshop on Intelligent Agents for Telecom Applications - ECAI'96*. 1996.

[312] P. Sousa. Interoperability of Networks for Interoperable Services. *IEEE Communications Magazine*, 37(5), May 1999.

[313] IPv6 Related Specifications. http://www.ipv6.org/.

[314] G. D. Stamoulis, D. Kalopsikakis, and A. Kirikoglou. Efficient-Based Negotiation for Telecommunication Services. In *Proceedings of GLOBECOM'99, Global Telecommunication Conference*, pages 1989–1996, 1999.

[315] T. Sugawara and K. Murakami. A multi-agent diagnostic system for internetwork problems. In *Proc. INET'92.*, 1992.

[316] R. Sun, B-T. Chu, R. Wilhelm, and J. Yao. A csp-based model for integrated supply chains. In *Working Notes of the Workshop: Artificial Intelligence for Electronic Commerce*, Orlando, Florida, July 1999.

[317] 3Com's switches. http://www.3com.com/products/switches.html.

[318] K. Sycara. Persuasive argumentation in negotiation. In *Theory and Decision*, pages 28:203–242. 1990.

[319] K. P. Sycara. Multiagent compromise via negotiation. In L. Gasser and M. N. Huhns, editors, *Distributed Artificial Intelligence, Volume II*, pages 119–137. Morgan Kaufmann, San Mateo, California, 1989.

[320] Information Society Technology. http://www.cordis.lu/ist/projects.htm.

[321] B. Teitelbaum. Internet2 qbone: Building a testbed for differentiated services.

[322] David L. Tennenhouse, Jonathan M. Smith, W. David Sincoskie, David J. Wetherall, and Gary J. Minden. A survey of active network research. *IEEE Communications*, 35(1):80–86, January 1997.

[323] TINA. Distributed processing environment (tina-dpe), December 1995.

[324] TINA-C. http://www.tinac.com.

[325] TMForum. http://www.tmforum.org.

[326] I. Union. Visual telephone systems and equipment for local area networks which provide a non-guaranteed quality of service, 1996.

[327] International Telecommunications Union. http://www.itu.int.

[328] A. Unruh and P. Rosenbloom. Abstraction in problem solving and learning. In *Proceedings of the Eleventh International Joint ConferenceonArtificial Intelligence, pages 681–687, Detroit, MI, 1989.*, 1989.

[329] W. van der Linden and A. Verbeek. Coalition formation: a gametheoretic approach. In *In H. A. M. Wilke, editor, Coalition Formation, volume 24 of Advances in Psychology. North Holland, 1985.* 1985.

[330] Hal R. Varian. *Microeconomic Analysis.* W. W. Norton and Co., New York, NY, USA, 1992.

[331] Dinesh Verma. *Supporting Service Level Agreements on IP Networks.* Macmillan Technical Publisher, September 1999.

[332] VITAL. http://www.infowin.org/acts/rus/projects/ac003.htm.

[333] J. von Neumann and O. Morgenstern. Theory of games and economic behavior. 1947.

[334] T. Von-Ungern-Sternberg. Swiss auctions. *Econometrica*, 58:341–357, 1991.

[335] N. Vulkan and N. R. Jennings. Efficient mechanisms for the supply of services in multi-agent environments. In *Proc of 1st Int Conference on Information and Computation Economies*, pages 1–10, Charlestown, South Carolina, 1999.

[336] V. Wade, D. Lewis, M. Sheppard, and M. Tschichholz. A methodology for developing integrated multi-domain service management systems. *Lecture Notes in Computer Science*, 1238, 1997.

[337] N. Wakamiya, Murata M., and Miyahara H. Utility-based bandiwdth allocation scheme for real-time multimedia stream transfer. In *Proceedings of Second IFIP/IEEE International Conference on Management of Multimedia Networks and Services '98*, October 1998.

[338] Mark Wallace. Practical applications of constraint programming. *Constraints*, 1(1/2):139–168, 1996.

[339] Jean Walrand and Pravin Varaiya. *High-Performance Communication Networks.* Morgan Kaufmann Publishers, Inc., San Francisco, 1996.

[340] D.N. Walton and E.C. Krabbe. *Commitments in Dialogue: Basic Concepts of Interpersonal Reasoning.* State University of New York Press, New York, 1995.

[341] Q. Wang and J. Peha. State-dependent pricing and its economic implications. In *Proc. of the 7th International Conf. on Telecommunication Systems Modeling and Analysis*, March 1999.

[342] Z. Wang and J. Crowcroft. Quality-of-Service Routing for Supporting Multimedia Applications. *IEEE Journal on Selected Areas in Communications*, 14(7), 1996.

[343] R. Weihmayer and R. Brandau. Cooperative distributed problem solving for communication network management. pages 547–557, 1990.

[344] R. Weihmayer and H. Velthuijsen. Intelligent Agents in Telecommunications. In N. R. Jennings and M. Wooldridge, editors, *Agent Technology Foundations, Applications and Markets*, pages 201–217. Springer Verlag and UNICOM UK, 1998.

[345] G. Weiss. *A Modern Approach to Distributed Artificial Intelligence.* MIT Press, 1999.

[346] G. Weiss. *Multiagent Systems.* MIT Press, 2000.

[347] D. S Weld. The use of aggregation in causal simulation. *Artificial Intelligence,* 30(1):1–34, 1986.

[348] Daniel Weld and Sanjaya Addanki. Task-driven model abstraction. presented at 4th International Workshop on Qualitative Physics, Lugano, Switzerland, 1990.

[349] M. Wellman. A market-oriented programming environment and its application to distributed multicommodity flow problems. *Journal of Artificial Intelligence,* pages 123–, 1993.

[350] M. P. Wellman. A Market-Oriented Programming Environment and its Application to Distributed Multicommodity Flow Problems. *Journal of Artificial Intelligence Research,* 1(1):1–23, 1994.

[351] S. Willmott and M. Calisti. An Agent Future for Network Control? *Swiss Journal of Computer Science (Informatik/Informatique),* 2000(1), January 2000.

[352] S. Willmott and B. Faltings. Bandwidth Adaptive Hierarchies for On-Line ATM Quality of Service Routing. In *Proceedings of the Conference on High Performance Routing and Switching: joint IEEE ATM Workshop 2000 and 3rd International Conference on ATM,* page tba. 2000.

[353] S. Willmott and B. Faltings. The Benefits of Environment Adaptive Organisations for Agent Coordination and Network Routing Problems. In *Proceedings of the Fourth International Conference on Multi Agent Systems (ICMAS-2000),* page tba. IEEE Press (in print), 2000.

[354] S. N. Willmott and B. Faltings. Active Organisations for Routing. In S. Covaci, editor, *Proceedings of the First International Working Conference on Active Networks.,* pages 262–273. Springer Verlag, Lecture Notes in Computer Science Series, number 1653, 1999.

[355] M. Wooldridge. Agent-based software engineering. *IEE Proceedings Software Engineering,* 144(1):26–37, 1997.

[356] M. Wooldridge and N.R. Jennings. Intelligent agents: Theory and practice. *The Knowledge Engineering Review,* 10(2):115–152, 1995.

[357] Michael Wooldridge and Nicholas R. Jennings. Pitfalls of agent-oriented development. In Katia P. Sycara and Michael Wooldridge, editors, *Proceedings of the 2nd International Conference on Autonomous Agents (Agents'98),* pages 385–391, New York, 9–13, 1998. ACM Press.

[358] CCITT Recommendation X.25. Interface between data terminal equipment, 1989.

[359] CCITT Recommendation X.711. Data communication networks- open systems interconnection. 1991.

[360] Fipa XC00029E. Fipa contract net interaction protocol specification. *Foundation for Intelligent Physical Agents,* 2000.

[361] Fipa XC00030E. Fipa iterated contract net interaction protocol specification. *Foundation for Intelligent Physical Agents,* 2000.

[362] X. Xiao, A. Hannan, and B. Bailey. Traffic engineering with mpls in the internet. 2000.

[363] X. Xiao, T. Irpan, A. Hannan, and R. C. Tsay. Traffic Engineering with MPLS. *America's Network Magazine*, pages 32–37, 1999.

[364] Xipeng Xiao. *Providing Quality of Service in the Internet*. PhD thesis, Miching State University, 2000.

[365] H. Yamaki, M. P. Wellman, and T. Ishida. A Market-Based Approach to Allocating QoS for Multimedia Applications. In *Proceedings: International Conference on Multi Agent Systems*, pages 385–392. 1996.

[366] F. Ygge. *Market-Oriented Programming and its Application to Power Load Management*. PhD thesis, Lund University, Lund, Schweden,, 1998.

[367] Fredrik Ygge and Hans Akkermans. Power load management as a computational market. In Victor Lesser, editor, *Proceedings of the First International Conference on Multi–Agent Systems*. MIT Press, 1995.

[368] M. Yokoo, E. H. Durfee, T. Ishida, and K. Kuwabara. Distributed Constraint Satisfaction for Formalising Distributed Problem Solving. *Proceedings 12th IEEE International Conference on Distributed Computing Systems.*, pages 614–621, 1992.

[369] M. Yokoo, E. H. Durfee, T. Ishida, and K. Kuwabara. The distributed constraint satisfaction problem: Formalization and algorithms. *IEEE Transactions on Knowledge and Data Engineering*, 10(5):673–685, September/October 1998.

[370] M. Yokoo and K. Hirayama. Distributed Breakout Algorithm for Solving Distributed Constraint Satisfaction Problems. In *Proceedings of the second International Conference on Multi Agent Systems (ICMAS'96)*, pages 401–408. 1996.

[371] M. Yokoo and K. Hirayama. Frequency assignment for cellular mobile systems using constraint satisfaction techniques. *Proceedings of the IEEE Annual Vehicular Technology Conference*, 2000.

[372] Makoto Yokoo. Asynchronous weak-commitment search for solving large-scale distributed constraint satisfaction problems. In Victor Lesser, editor, *Proceedings of the First International Conference on Multi–Agent Systems*, page 467, San Francisco, CA, 1995. MIT Press. (poster).

[373] Makoto Yokoo and Katsutoshi Hirayama. Distributed breakout algorithm for solving distributed constraint satisfaction problems. In Victor Lesser, editor, *Proceedings of the First International Conference on Multi-Agent Systems*. MIT Press, 1995.

[374] Makoto Yokoo and Katsutoshi Hirayama. Algorithms for distributed constraint satisfaction: A review. *Autonomous Agents and Multi-Agent Systems*, 3(2):185–207, June 2000.

[375] Dajun Zeng and Katia Sycara. Benefits of learning in negotiation. In *Proceedings of AAAI-97*, 1997.

[376] G. Zlotkin and J. S. Rosenschein. A domain theory for task oriented negotiation. In A. Cesta, R. Conte, and M. Micheli, editors, *Pre-Proceedings of MAAMAW-92*, July 1992.